Advances in Geophysical
Mechanics and Mathema

AF213024

Series editors

Angelika Humbert, Bremerhaven, Germany
Kolumban Hutter, Zürich, Switzerland

For further volumes:
http://www.springer.com/series/7540

Chyan-Deng Jan

Gradually-varied Flow Profiles in Open Channels

Analytical Solutions by Using Gaussian Hypergeometric Function

Springer

Chyan-Deng Jan
Department of Hydraulic and Ocean
Engineering
National Cheng Kung University
Tainan
Taiwan

ISSN 1866-8348 ISSN 1866-8356 (electronic)
ISBN 978-3-662-51168-8 ISBN 978-3-642-35242-3 (eBook)
DOI 10.1007/978-3-642-35242-3
Springer Heidelberg New York Dordrecht London

© Springer-Verlag Berlin Heidelberg 2014
Softcover reprint of the hardcover 1st edition 2014
This work is subject to copyright. All rights are reserved by the Publisher, whether the whole or part of the material is concerned, specifically the rights of translation, reprinting, reuse of illustrations, recitation, broadcasting, reproduction on microfilms or in any other physical way, and transmission or information storage and retrieval, electronic adaptation, computer software, or by similar or dissimilar methodology now known or hereafter developed. Exempted from this legal reservation are brief excerpts in connection with reviews or scholarly analysis or material supplied specifically for the purpose of being entered and executed on a computer system, for exclusive use by the purchaser of the work. Duplication of this publication or parts thereof is permitted only under the provisions of the Copyright Law of the Publisher's location, in its current version, and permission for use must always be obtained from Springer. Permissions for use may be obtained through RightsLink at the Copyright Clearance Center. Violations are liable to prosecution under the respective Copyright Law.
The use of general descriptive names, registered names, trademarks, service marks, etc. in this publication does not imply, even in the absence of a specific statement, that such names are exempt from the relevant protective laws and regulations and therefore free for general use.
While the advice and information in this book are believed to be true and accurate at the date of publication, neither the authors nor the editors nor the publisher can accept any legal responsibility for any errors or omissions that may be made. The publisher makes no warranty, express or implied, with respect to the material contained herein.

Printed on acid-free paper

Springer is part of Springer Science+Business Media (www.springer.com)

To
Dr. Cheng-lung Chen

Preface

Many hydraulic engineering works involve the computation of surface profiles of gradually varied flow (GVF) that is a steady nonuniform flow in an open channel with gradual changes in its water surface elevation. The most widely used methods for computing GVF profiles could be classified into the direct integration methods, step methods (i.e., simple numerical methods), and advanced numerical methods (e.g., the standard fourth order Runge–Kutta method). Numerical solutions of the GVF equation are primarily used in non-prismatic channels. In some prismatic channels, such as artificial channels, the GVF equation can be simplified so as to let the analytical (or semi-analytical) direct integration be applied. The analytical direct-integration method is straightforward and can provide the total length of the profile in a single computation step.

In the direct-integration method, the one-dimensional GVF equation is usually normalized to a simpler expression in advance so as to allow the performance of direct integration. In most cases, the GVF equation is normalized by the normal depth, while in some cases, it is normalized by the critical depth. Many attempts have been made by previous investigators on the direct-integration method to solve the GVF equation. The table of the varied-flow function (VFF) is needed in the Bakhmeteff-Chow procedure to obtain the GVF profiles by using the conventional direct-integration method. The use of the conventional Bakhmeteff-Chow procedure to obtain GVF profiles has a drawback caused by the inconvenience of the use of the VFF-table and the imprecise interpolation of the VFF-values from the VFF table. To overcome the drawback, Dr. C. L. Chen and the author had initiated a self-investigation to look for an alternative method to get analytical solutions of GVF profiles without recourse to the VFF since 2005. Dr. Chen was an excellent hydrologist in the U.S. Geological Survey, and a distinguished professor in the University of Illinois at Urbana-Champaign and in the Utah State University. Unfortunately, Dr. Chen passed away in January 2012, just about the time at which we found that the Gaussian hypergeometric functions (GHF) can be used to analytically solve the GVF equation, and completed two drafts of manuscripts to be submitted to a suitable journal. After some revisions by the author, these two manuscripts were submitted to two journals and finally published, one in the Journal of Hydrology in August 2012, and another in the Hydrology and Earth System Sciences (HESS) in March 2013, respectively.

Professor K. Hutter, the editor of Advances in Geophysical and Environmental Mechanics and Mathematics (AGEM^2), encouraged the author to thoroughly write out the innovative method using the GHF in analytically solving the GVF profiles in the form of a book, when he visited our university in May 2012. The author began to prepare this book in September 2012. This book attempts to thoroughly introduce the innovative procedure to obtain the analytical solution of GVF profiles using the Gaussian hypergeometric functions (GHF) as well as to present the analysis and discussions of the GHF-based solutions. This book is divided into five chapters. Chapter 1 introduces the basic equations for the GVF in open channels. Chapter 2 introduces the conventional direct-integration methods used to analytically solve the GVF equation. Chapter 3 presents the GVF equation normalized by the normal depth, and the application of the GHF on analytically solving the normal-depth-based dimensionless equation for GVF in sustaining channels. The classification and properties of the GHF-based solutions of the GVF profiles are also discussed in this chapter. Chapter 4 is devoted to the GVF equation normalized by the critical depth, and then solves it by using GHF for GVF flows in sustaining and non-sustaining channels. Chapter 5 presents the analysis and classification of the GHF-based solutions of the critical-depth-based dimensionless GVF equation. This chapter also shows that the critical-depth-based GVF solution expressed in terms of the GHF is more useful and versatile than its counterpart normalized by using the normal depth.

The author is especially indebted to Prof. C. L. Chen who introduced the author to the subject of GVF profiles as presented in this book, as well as to the subject of debris flow when the author was a graduate student in the University of California, Berkeley between 1988 and 1992, and when the author was a visiting scholar at the U.S. Geological Survey at Menlo Park, California, in 1999. The author also acknowledges Prof. Y. C. Tai and Miss C. C. Shen for their help in the transformation of the manuscript typed in the form of Microsoft Word to that in the form of LaTex. Thanks are due to Prof. K. Hutter and Dr. A. Siviglia for their reviews and valuable suggestions. The author appreciates the supports and resources provided by the National Cheng Kung University, and the National Science Council, Taiwan. Finally, the author would like to express his sincere thanks to all those who have directly or indirectly helped him in writing this book, and to Springer Verlag and its personnel for their help in the production of the book.

The author sincerely hopes that this book will be a reference book for practical civil or hydraulic engineers when they design hydraulic engineering works as well as for undergraduate and graduate students in the fields of civil, hydraulic, and agricultural engineering.

Taiwan, September 1, 2013 Chyan-Deng Jan

Contents

Acronyms and Symbols

Symbol	Name/Descripition
A	Adverse-slope channel
A1, A2, A3	GVF profiles in adverse-slope channels
A, A_n	Cross-section areas
B	Bottom width
C	Critical-slope channel
C1, C2, C3	GVF profiles in critical-slope channels
C	Factors of the flow resistance
C_k	Coefficient relating K and h
C_z	Coefficient relating Z and h
D	Hydraulic depth
d_0	Diameter of a circular channel
ETF	Elementary transcendental function (s)
GVF	Gradually varied flow
GHF	Gaussian hypergeometric function (s)
F	Froude number
g	Gravitational acceleration
$g(b, z)$	Simplified notation of Gaussian hypergeometric function
H	Horizontal-bed channel
H1, H2, H3	GVF profiles in horizontal-bed channels
H	Total energy
h	Flow depth normal to the channel bed
h_c	Critical depth
h_e	Eddy loss
h_f	Frictional loss
h_n	Normal depth
J	Parameter, $J = N/(N - M + 1)$
K, K_c, K_n	Conveyances of the channel section
K_u	Curvature of a GVF profile
M	Mild-slope channel
M1, M2, M3	GVF profiles in mild-slope channels
M	Hydraulic exponent for critical-flow computation
m	Exponent used in the power law of the wall
N	Hydraulic exponent for uniform-flow computation

n	Manning's roughness coefficient
P	Wetted perimeter of cross-section
p	Exponent relating to hydraulic radius
Q	Flow discharged
q	Exponent relating to slopes
R, R_n	Hydraulic radiuses of flow
S	Steep-slope channel
S1, S2, S3	GVF profiles in steep-slope channels
S_0	Slope of channel bed ($= \sin\theta$)
S_{0_*}	$\tan\theta$
S_c	Critical slope
S_{c_*}	$S_c/\cos\theta$
S_f	Energy slope
T, T_c	Top widths of the flow cross-section
u	Dimensionless flow depth ($= h/h_n$)
VFF	Varied-flow function
V, V_n	Mean velocities of flow
v	Dimensionless flow depth ($\upsilon = h/h_c$ or $= u^{N/J}$)
w	The reciprocal of u or λv
x	Longitudinal coordinate along the channel bed
x_*	Dimensionless longitudinal coordinate ($= xS_{0_*}/h_n$)
$x_\sharp$	Dimensionless longitudinal coordinate ($= xS_{c_*}/h_c$)
y	Flow depth normal to the datum
Z, Z_c	Section factors
z	Channel side slope, or function variable
z_b	Elevation of channel bed above the datum
α	Energy coefficient
β	Momentum coefficient
θ	Inclined angle of channel bed
λ	Parameter ($= h_c//h_n$)
ψ	Parameter
$\Gamma(a)$	Gamma function

Chapter 1
Basic Equations for the Gradually-Varied Flow

1.1 Introduction

The one-dimensional gradually-varied flow (GVF) is a steady non-uniform flow in a prismatic channel with gradual changes in its water surface elevation. Many hydraulic engineering works involve the computation of one-dimensional GVF surface profiles such as the drawdown produced at a sudden drop in a channel and the backwater produced by a dam or weir across a river, as indicated by Chaudhry (2006), Chow (1959), Subramanya (2009), Jan and Chen (2012), Vatankhah (2012), among others. The evaluation of steady one-dimensional gradually-varied flow profiles under a specific flow discharge is very important in open-channel hydraulic engineering. Two basic assumptions are involved in the analysis of GVF. One is the pressure distribution at any section assumed to be hydrostatic. The other is the resistance to flow at any depth assumed to be given by the corresponding uniform flow equation, such as Mannings equation. Almost all major hydraulic-engineering activites in free surface flow involve the computation of GVF profiles. The various available procedures for computing GVF profiles can be classified as: the graphical-integration method, the direct integration, and the numerical method, as shown in Chow (1959) and Subramanya (2009). The development of the basic GVF dynamic equation and the classification of flow profiles in a prismatic channel is reviewed and discussed in this chapter. The direct integration method for analytically solving the GVF equation by using the Gaussian hypergeometric function (GHF) will be presented in subsequent chapters.

1.2 The GVF Equation for Flow in Open Channels

The GVF equation for flow in open channels with various cross-sectional shapes can be readily derived from the energy or momentum conservation law. Therefore, two types of the GVF equation are obtainable: One is based on the energy conservation law and the other on the momentum conservation law. The difference between them simply lies in the convective acceleration term expressed using the different

C.-D. Jan, *Gradually-varied Flow Profiles in Open Channels*,
Advances in Geophysical and Environmental Mechanics and Mathematics,
DOI: 10.1007/978-3-642-35242-3_1, © Springer-Verlag Berlin Heidelberg 2014

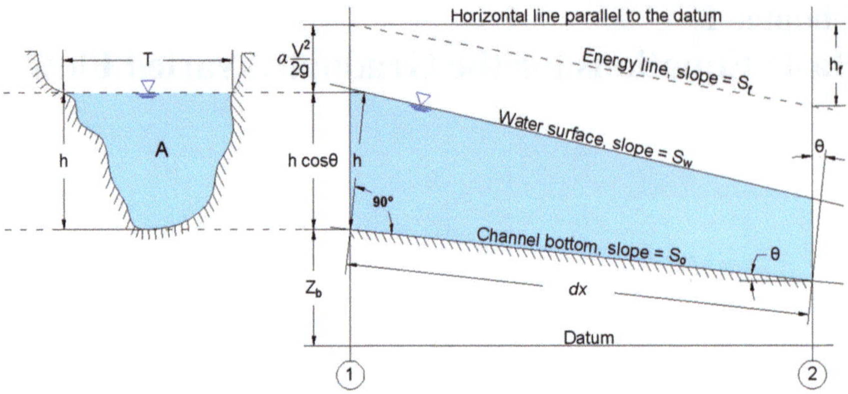

Fig. 1.1 Derivation of the gradually-varied flow equation

velocity-distribution correction factor (or coefficient), namely the energy coefficient, α, and the momentum coefficient, β, based on the energy and momentum conservation laws, respectively. The theoretical expressions of α and β for turbulent shear flow in wide channels can be derived from the power law, and the exponent (m) of the power-law velocity distribution is the sole parameter that determines the values of α and β, as indicated by Chen (1992). Many investigators, such as Chow (1959), have adopted α for its wide recognition and comprehension. Thus, following in most researchers footsteps, we opt to use α rather than β. Nevertheless, before elaborating on how to solve the GVF equation, we briefly revisit the derivation of the GVF equation based on the energy conservation law. We proceed to express the GVF equation in terms of the flow depth, h, only. We further normalize it using the characteristic length, such as the normal depth, h_n, as shown in Chaps. 2 and 3 or the critical depth, h_c, as shown in Chaps. 4 and 5.

To derive the GVF equation, we consider a GVF profile in an open prismatic channel with arbitrary cross-sectional shape. For simplicity, let us assume neither vertical acceleration nor departure from the hydrostatic pressure distribution in flow. Then, as shown in Fig. 1.1, the total energy H of a gradually-varied flow at any section referred to a datum in a channel can be written as

$$H = z_b + h\cos\theta + \frac{\alpha V^2}{2g}, \tag{1.1}$$

in which z_b is the vertical coordinate of the channel bottom above the datum, h is the flow depth measured perpendicularly from the channel bottom, θ is the angle of the channel-bottom slope, V is the mean velocity of flow through the section, and g is the gravitational acceleration. For small slope angle θ, $\cos\theta \approx 1.0$, (1.1) becomes

$$H = z_b + h + \frac{\alpha V^2}{2g}. \tag{1.2}$$

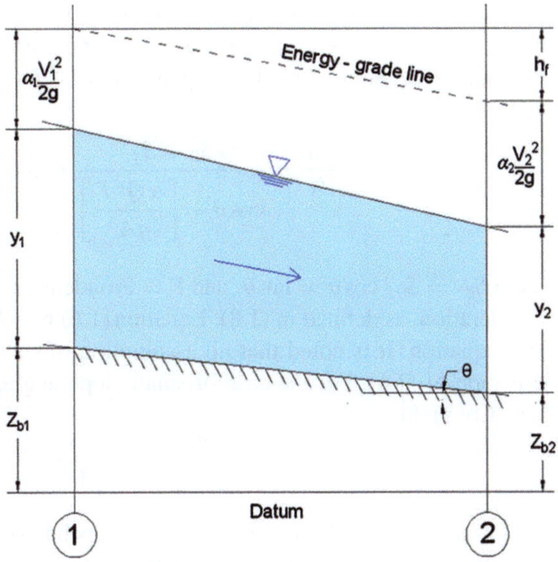

Fig. 1.2 Energy relation for a gradually-varied flow in channels of small slope

Since the channel slope is small in most practical problems, many researchers even use the flow depth y measured in the direction normal to the datum (as shown in Fig. 1.2) to replace h or $h\cos\theta$, i.e.,

$$H = z_b + y + \frac{\alpha V^2}{2g}. \tag{1.3}$$

For uniform flow, the relation between h and y is given by $h = y\cos\theta$. Considering the effect of channel slope and keeping the original meaning of flow depth measured normally to the bed, we will use (1.1) in the subsequent discussions. Taking the channel bottom as the longitudinal coordinate x and differentiating (1.1) with respect to x on the assumption that θ and α do not vary with x yields

$$\frac{dH}{dx} = \frac{dz_b}{dx} + \cos\theta\frac{dh}{dx} + \alpha\frac{d}{dx}\left(\frac{V^2}{2g}\right), \tag{1.4}$$

where $dH/dx(= -S_f)$ is the energy slope (namely, friction slope), $dz_b/dx(= -S_0)$ is the bottom slope, $S_0 = \sin\theta$, and dh/dx is the water surface slope relative to the bottom of the channel. The mean velocity can be written as $V = Q/A$, in which Q is the flow discharge and A is the area of flow cross-section. The last term of (1.4) can be written as

$$\alpha\frac{d}{dx}\left(\frac{V^2}{2g}\right) = \frac{\alpha}{2g}\frac{d}{dx}\left(\frac{Q^2}{A^2}\right) = -\frac{\alpha Q^2}{gA^3}\frac{dA}{dh}\frac{dh}{dx} = -\left(\frac{\alpha Q^2 T}{gA^3}\right)\frac{dh}{dx}, \tag{1.5}$$

where $T = (dA/dh)$ is the top width of the flow cross-section at the free surface, as shown in Fig. 1.1. Therefore, the basic governing equation of GVF (also known as the dynamic equation of GVF) in open channels can be written in the form

$$\frac{dh}{dx} = \frac{S_0 - S_f}{\cos\theta - \left[\frac{\alpha Q^2 T}{g A^3}\right]} = S_{0*} \frac{1 - \frac{S_f}{S_0}}{1 - \mathbf{F}^2}, \tag{1.6}$$

where $S_{0*} = S_0/\cos\theta = \tan\theta$, and $\mathbf{F}$ = Froude number for GVF at a section under consideration, as defined in (1.8). Equation (1.6) may be called the one-dimentional GVF equation. It is noted that no assumption of small θ has been imposed in the derivation of (1.6). For the case of small slope angle, $\cos\theta \approx 1.0$, $S_{0*} \approx S_0$, and then (1.6) yields

$$\frac{dh}{dx} = \frac{S_0 - S_f}{1 - \mathbf{F}^2}. \tag{1.7}$$

The Froude number $\mathbf{F}$ in (1.6) and (1.7) depends on the flow discharge, channel geometry as well as channel slope, and is given by the relation

$$\mathbf{F} = \sqrt{\frac{\alpha Q^2 T}{g A^3 \cos\theta}} = \frac{V}{\sqrt{g D \cos\theta/\alpha}}, \tag{1.8}$$

in which $D(= A/T)$ is the hydraulic depth. The Froude number, as defined above, varies with x, and the effects of α and θ are also taken into account in defining $\mathbf{F}$. The flow is subcritical if $\mathbf{F} < 1$, and supercritical if $\mathbf{F} > 1$. When $\mathbf{F} = 1$, the flow is critical and the water surface slope tends to infinity, and thus the surface profile approaches the critical-depth line vertically because $dh/dx \to \infty$. For uniform flow, $S_0 = S_f$ and dh/dx tends to zero, and thus the surface profile approaches the normal-depth line asymptotically. Without consideration of the effects of α and θ, the definition of $\mathbf{F}$ can be written as

$$\mathbf{F} = \sqrt{\frac{Q^2 T}{g A^3}} = \frac{V}{\sqrt{g D}}. \tag{1.9}$$

1.3 The GVF Equation in Terms of Flow Depth

1.3.1 Conveyance and Section Factor of Channel Section

To express (1.6) in terms only of h, we first denote the conveyance of the channel section as K $(= Q/\sqrt{S_f})$ and its counterpart for the corresponding uniform flow of a given discharge Q at the normal depth, h_n, as $K_n(= Q/\sqrt{S_0})$. Then, we denote the section factor for flow as Z and its counterpart for the corresponding critical flow of Q at the critical depth, h_c, as Z_c. Assuming that Q is constant for GVF in the channel, we can respectively express the second term in both numerator and denominator on the right-hand side of the last equal sign in (1.6) as

$$\frac{S_f}{S_0} = \left(\frac{K_n}{K}\right)^2, \tag{1.10}$$

$$\mathbf{F}^2 = \frac{\alpha Q^2 T}{gA^3 \cos\theta} = \frac{\alpha Q^2}{g \cos\theta \left(A^2 D\right)} = \left(\frac{Z_c}{Z}\right)^2, \tag{1.11}$$

in which $K = CAR^p$, $K_n = CA_n R_n^p$, $Z^2 = A^2 D$, $Z_c^2 = \alpha Q^2 / g \cos\theta$ and $D = A/T$ = hydraulic depth. The additional symbols used in the preceding notation are defined as follows: C = factor of the flow resistance (e.g., Chézy's resistance factor if the Chézy formula is used or the reciprocal of Manning's n if the Manning formula in SI units is used), A = cross-sectional area of flow, $R = A/P$ = hydraulic radius of flow, P = wetted perimeter, p = exponent of R or R_n as related to the mth power of the power-law flow resistance formula, as shown by Chen (1991), T = top width of free surface, and symbols denoted by subscripts n and c represent that entity pertaining to "normal flow" and "critical flow", respectively.

Chézy formula:

$$Q = CA_n R_n^{1/2} S_0^{1/2} \quad \text{(for uniform flow)}, \tag{1.12}$$

$$Q = CAR^{1/2} S_f^{1/2} \quad \text{(for gradually-varied flow)}; \tag{1.13}$$

Manning formula (in SI units):

$$Q = \frac{1}{n} A_n R_n^{2/3} S_0^{1/2} \quad \text{(for uniform flow)}, \tag{1.14}$$

$$Q = \frac{1}{n} AR^{2/3} S_f^{1/2} \quad \text{(for gradually-varied flow)}. \tag{1.15}$$

Before rearranging (1.6) in a form expressible in terms only of the flow depth, h, from which we seek a solution for GVF profiles in channels, one must be aware of the common notion implied in the formulation of (1.6), namely a tacit assumption made on the validity of (1.6). It is imposed on a flow condition under which the Froude number, $\mathbf{F}$, for flow at any section in a channel under study is less than the stability limit, so that the flow cannot generate free-surface instability (Chen 1995), such as roll waves. Kranenburg (1990) ever discussed the stability criteria and growth rates of long-wave disturbances in gradually varying open-channel flow. Without this tacit assumption imposed on (1.6), GVF profiles to be computed from (1.6) may not be warranted. Equation (1.6) on substitution of (1.10) and (1.11) in virtue of $S_{0*} = \tan\theta$ yields

$$\frac{dh}{dx} = S_{0*} \frac{1 - (K_n/K)^2}{1 - (Z_c/Z)^2}, \tag{1.16}$$

which should be identical to its counterpart derived from the momentum conservation law because neither α nor β appears in (1.16) owing to its embedment in the expression of Z_c. In addition, the water depth calculated from the equation of uniform flow, such as Chézy formula or Manning formula, is called as the normal depth (h_n). Similary, the water depth calculated from the critical-flow condition, i.e., the Froude number $\mathbf{F} = 1.0$ is called as the critical depth (h_c). Both h_n and h_c are two of the most important parameters in hydraulics, especially in the classification of GVF profiles.

1.3.2 Hydraulic Exponents Defined in Relation to Conveyance and Section Factor

The hydraulic exponents, M and N, for critical-flow and normal-flow computations are, respectively, defined in relation to the section factors, Z and Z_c, and the conveyances, K and K_n, as follows:

$$Z^2 \left(= \frac{A^3}{T} \right) = C_z h^M, \tag{1.17}$$

$$Z_c^2 \left(= \frac{A_c^3}{T_c} = \frac{\alpha}{\cos\theta} \frac{Q^2}{g} \right) = C_z h_c^M, \tag{1.18}$$

$$K^2 \left[= (CAR^p)^2 = \frac{Q^2}{S_f^{2q}} \right] = C_k h^N, \tag{1.19}$$

$$K_n^2 \left[= (CA_n R_n^p)^2 = \frac{Q^2}{S_0^{2q}} \right] = C_k h_n^N. \tag{1.20}$$

in which C_z = coefficient relating Z with h (or Z_c with h_c), C_k = coefficient relating K with h (or K_n with h_n), M = hydraulic exponent for critical-flow computation, and N = hydraulic exponent for uniform-flow computation, as defined earlier. It is noted in the formulation of (1.17) through (1.20) that the section factor (Z or Z_c) and the conveyance (K or K_n) are expressed in terms only of h with the two exponents, M and N, which themselves are in fact also functions of h (or h_c or h_n) as pointed out by Chow (1959, 1981). Meriting mention is the factor of the flow resistance, C, which varies with channel roughness and viscosity, among other factors, as manifesting itself in the power-law flow resistance formula (Chen 1991).

Next, attention is called to the implication of (1.16), which has been derived without taking into account the exponent, q, of the friction slope, S_f, as well as the channel-bottom slope, S_0, as embraced in the power-law flow resistance formula (Chen 1991). Before adjusting (1.16) by taking q into account, we remark the relations within the parentheses in (1.17) and (1.18) as well as the relations within the brackets in (1.19) and (1.20). Because the flow under consideration is not necessarily at the critical state, we can readily establish the following inequality.

$$\frac{A^3}{T} \neq \frac{\alpha}{\cos\theta}\frac{Q^2}{g}. \tag{1.21}$$

On the contrary, the relation within the parentheses in (1.18) can proved to be true because the flow under consideration is at the critical state. Accordingly, combining (1.17) and (1.18) with the help of $D = A/T$ yields

$$\left(\frac{Z_c}{Z}\right)^2 = \left(\frac{\alpha}{\cos\theta}\frac{Q^2}{gA^2D}\right) = \left(\frac{h_c}{h}\right)^M, \tag{1.22}$$

which essentially gives the same relation shown in (1.11). By contrast, the expression of $(K_n/K)^2$ in (1.10) or (1.16) should be adjusted for the exponent, q, of the friction slope, S_f, and the channel-bottom slope, S_0, in addition to the exponent, p, of R (or R_n) in the power-law flow resistance formula, i.e.,

$$V = CR^pS_f^q \tag{1.23}$$

or

$$V_n = CR_n^pS_0^q, \tag{1.24}$$

in which $V_n(= Q/A_n)$ is the mean velocity of the section for the corresponding uniform flow. Recalling that q is not necessarily equal to 1/2 (Chow 1959; Chen 1991), even though $q = 1/2$ is always adopted in the Chézy formula or the Manning formula, as showed in (1.12)–(1.15), we incorporate (1.19) in (1.20) to express

$$\left(\frac{K_n}{K}\right)^2 = \left(\frac{S_f}{S_0}\right)^{2q} = \left(\frac{h_n}{h}\right)^N \tag{1.25}$$

or

$$\left(\frac{K_n}{K}\right)^{1/q} = \frac{S_f}{S_0} = \left(\frac{h_n}{h}\right)^{N/2q}. \tag{1.26}$$

Because (1.16) has been obtained from (1.6) on substitution of S_f/S_0 using (1.10), the adjustment of (1.16) is required by replacing its $(K_n/K)^2$ by $(K_n/K)^{1/q}$, i.e.,

$$\frac{dh}{dx} = S_{0*}\frac{1-(K_n/K)^{1/q}}{1-(Z_c/Z)^2} \tag{1.27}$$

or, if expressed in terms of the flow depth, (1.16) can be further reconciled in the light of (1.22) and (1.25) to

$$\frac{dh}{dx} = S_{0*} \frac{1 - (h_n/h)^{N/2q}}{1 - (h_c/h)^M}. \tag{1.28}$$

Equation (1.28) is a reconciled form of the GVF equation expressed in terms only of the flow depth, h. It is well known that the hydraulic exponents, M and N, are the functions not only of h but also of channel geometry (Chow 1959) and channel roughness, which involves a parameter, m, i.e., the exponent in the power-law flow resistance formula (Chen 1991). Being overwhelmed by the difficulty in integrating (1.28) exactly over h for the general case of a channel with regular channel geometry such as rectangle, trapezoid, triangle, and circle, among others, we only attempt herein to deal with the simplest case of a wide channel in which $M = 3$. In other words, with the present knowledge in mathematics, it is unlikely that we can solve (1.28) analytically if M and N are functions of h. For simplicity, we assume that M and N are constant hereafter in this book, thereby expressing N (but not M) in term of the m-value via the exponent, p, of R in (1.23). Thus, given the m-value, how we determine the N-value from it is an important issue yet to be resolved judiciously in the following before we can integrate (1.28) exactly over h.

1.3.3 Role of the Power-Law Flow Resistance Formula in the GVF Equation

By and large, the power-law flow resistance formulas, as shown generally in (1.23) and (1.24), can be classified into two types: One for hydraulically smooth flows and the other for fully rough flows (Chen 1991). We may expressed the former in the form of (1.23) and (1.24) as

$$V = CR^{(3m+1)/2} S_f^{(m+1)/2} \tag{1.29}$$

or

$$V_n = CR_n^{(3m+1)/2} S_0^{(m+1)/2} \tag{1.30}$$

and the latter as

$$V = CR^{(2m+1)/2} S_f^{1/2} \tag{1.31}$$

or

$$V_n = CR_n^{(2m+1)/2} S_0^{1/2}, \tag{1.32}$$

in which m is the exponent of a unified similarity variable (dimensionless) used in the power laws of the wall. Comparing each variable in (1.23) and (1.24) with its counterpart in (1.29) and (1.30) or (1.31) and (1.32), we can establish the following relations between the exponents of the corresponding variables. For hydraulically smooth flows, we have

$$p = \frac{3m + 1}{2} \tag{1.33}$$

and

$$q = \frac{m + 1}{2}. \tag{1.34}$$

For fully rough flows,

$$p = \frac{2m + 1}{2} \tag{1.35}$$

and

$$q = \frac{1}{2}. \tag{1.36}$$

Equations (1.33)–(1.36), which express the values of p and q in terms of the m-value, can be used to determine the N-value for hydraulically smooth flows and fully rough flows, respectively. Although both equations are applicable for flow in channels with any cross-sectional shape, we use them only for flow in wide channels herein. Because $A = h$ (or $A_c = h_c$) and $T = 1$ (or $T_c = 1$) for flow in wide channels, we can readily establish $M = 3$ from (1.17) and (1.18) by virtue of (1.25) and (1.26). By the same token, because $A = h$ (or $A_n = h_n$) and $R = h$ (or $R_n = h_n$) for flow in wide channels, we can equate the corresponding exponents of h (or h_n) on both sides of the equal sign in (1.19) and (1.20), or (1.25), thereby establishing another relation as follows:

$$\frac{1 + p}{q} = \frac{N}{2q} \tag{1.37}$$

or

$$N = 2(1 + p). \tag{1.38}$$

Equation (1.38) on substitution of p from (1.33) and (1.35) yields respectively for hydraulically smooth flows

$$N = 3(m + 1) \tag{1.39}$$

and for fully rough flows

$$N = 2m + 3. \tag{1.40}$$

The N-value for flow in wide channels can thus be determined from (1.39) or (1.40), provided that the m-value for hydraulically smooth flows or fully rough flows is given or known.

Given $M = 3$ for flow in wide channels, we evaluate the modified exponent, $N/2q$, of h_n/h appearing in the numerator of (1.28), i.e., a reconciled form of the GVF equation, before we solve (1.28) analytically for h. Evaluation of $N/2q$ for smooth and rough walls can proceed differently as follows: For hydraulically smooth flows in wide channels, the magnitude of $N/2q$ upon substitution of the expressions of N and q from (1.39) and (1.34), respectively, yields 3, which is independent of the m-value. In other words, the value of $N/2q$ for hydraulically smooth flows in wide channels is equivalent to the N-value obtained based on the Chézy formula, namely $N = 3$, irrespective of its m-value. On the other hand, for fully rough flows in wide channels, the expression of $N/2q$ upon substitution of N and $2q$ from (1.40) and (1.36), respectively, results in the very same expression of N, as shown in (1.40), because $2q = 1$.

In view of the aforementioned implication of the modified hydraulic exponent, $N/2q$, used in (1.28) for both hydraulically smooth flows and fully rough flows in wide channels, we prefer to express the GVF equation in terms of the flow depth, h, with the exponent, N, as

$$\frac{dh}{dx} = S_{0*}\frac{1-(h_n/h)^N}{1-(h_c/h)^M}, \tag{1.41}$$

rather than with the modified exponent, $N/2q$, as shown in (1.28). Therefore, if (1.41) is adopted, we use $N = 3$ and $N = 2m + 3$ for hydraulically smooth flows and fully rough flows in wide channels, respectively. Specifically, if $m = 0$ for fully rough flows, $N = 2m + 3$ on substitution of $m = 0$ reduces to 3, which is identical to the N-value obtained based on the Chézy formula for hydraulically smooth flows. Consequently, as far as flows in wide channels are concerned, we can henceforth use (1.41) in search of its solution for both hydraulically smooth and fully rough flows because the solution of (1.37) for $m = 0$ (i.e., $N = 3$) for fully rough flows can be regarded as that for any m-value for hydraulically smooth flows. This notion remains to be justified when (1.41) is later solved for an N-value equal to any real number. This also explains why the Chézy formula has been extensively used in the GVF computation in wide channels, regardless of whether flows under study are hydraulically smooth or fully rough, for $N = 3$ is applicable to both hydraulically smooth flows and fully rough flows if the Chézy formula is used in the formulation of (1.41). In open-channel hydraulics, M and N are usually called the hydraulic exponents for critical-flow computation and uniform-flow computation, respectively. Both of M and N are positive real numbers, depending on the shape of channel cross section as well as flow depth. Equation (1.41) is very helpful in understanding the meanings of classification of the gradually-varied flow profiles, as indicated in next section.

1.4 The GVF Equation for Flow in Adverse Channels

Considering the flow in adverse channels, we deem the slope of the channel bottom as negative, i.e., $S_0 < 0$, in the derivation of GVF equation for flow in adverse channels. Such a treatment may result in an imaginary value of K_n or a negative value of K_n^2 for

Table 1.1 Slope-based classification of channels

Channel category	Symbol	Characteristic consition	Remark
1. Horizontal bed	H	$S_0 = 0$	Cannot sustain uniform flow
2. Mild slope	M	$h_n > h_c$	Subcritical flow at normal depth
3. Critical slope	C	$h_n = h_c$	Critical flow at normal depth
4. Steep slope	S	$h_n < h_c$	Supercritical flow at normal depth
5. Adverse slope	A	$S_0 < 0$	Cannot sustain uniform flow

a given Q $(= K_n S_0^q$, provided $q = 1/2)$ as shown in Chow (1959). Although h_n for uniform flow in adverse channels is undefined, we may assume that the coefficient of the flow resistance, C, in the power-law flow resistance formula is imaginary so as to enable the computation of h_n (Chen 1991). In other words, for a negative value of K_n^2, the value of C^2 used in computing h_n from $K_n^2 = C^2 A_n^2 R_n^{2p}$ (where p is the exponent of R_n as related to the mth power of the power-law flow resistance formula) must be negative, thus resulting in a fictitious value of h_n, if determined from the above relation.

Following the similar derivation process as indicated in the previous sections for flow in sustaining channels, we can obtain the GVF equation for flow in adverse channels in terms of flow depth h as

$$\frac{dh}{dx} = S_{0*} \frac{1 + (h_n/h)^N}{1 - (h_c/h)^M}, \tag{1.42}$$

in which S_{0*} $(= S_0/\cos\theta)$ is negative (i.e., $S_{0*} < 0$) because S_0 is negative. Equation (1.42) will be used to establish a h_c-based dimensionless GVF equation for flow in adverse channels, as shown in Chap. 4 (Sect. 4.2.3). The solutions of the dimensionless GVF equation for flow in adverse channels will be presented in Chaps. 4 and 5 in detail.

1.5 Classification of Gradually-varied Flow Profiles

If the flow discharge Q, the resistance coefficient (such as Manning's n), and the bottom slope S_0 are fixed in a given channel, the normal depth h_n and the critical depth h_c are two fixed depths. There are three possible relations between h_n and h_c as: (1) $h_n > h_c$, (2) $h_n < h_c$, (3) $h_n = h_c$. Further, there are two cases where h_n does not exist: one is for flow in a channel of a horizontal bed (where $h_n \to \infty$), and the other is for flow in a channel of an adverse bed slope (where S_0 is negative and h_n is virtual). Therefore, the channels can be classified into five categories: horizontal bed (H), mild slope (M), critical slope (C), steep slope (S), and adverse slope (A), as indicated in Table 1.1.

Lines representing the critical depth and normal depth (if it exists) can be drawn in the longitudinal section for each of the five slope-based categories of channels. These lines would divide the whole flow space into three zones as: Zone 1 where space is above the upper line; Zone 2 where space locates between the upper line and the next lower line; Zone 3 where space locates between the lower line and the bed. Further, to assist in the determination of GVF profiles in various regions, the behavior of the water surface slope (dh/dx) at certain key depths is noted from Eq. (1.41) and this information is usefully as indicative of the trend of the water surface profile as follows:

1. As $h \to h_n, dh/dx \to 0$, which indicates that the water surface asymptotically approaches the normal depth line.
2. As $h \to h_c, dh/dx \to \infty$, which indicates that the water surface vertically meets the normal depth line.
3. As $h \to \infty, dh/dx \to S_0$, which indicates that the water surface at a very large depth gradually meets a horizontal asymptote that is parallel to the line of datum.

The flow profiles may be classified into thirteen different types according to the nature of the channel slope and the zone in which the flow surface lies. These types are designed as H2, H3, M1, M2, M3, C1, C2, C3, S1, S2, S3, A2 and A3; where the letters are descriptive of the characteristic of channel slope: H for horizontal slope, M for mild slope critical slope, S for steep slope, and A for adverse slope, respectively; while the numerals 1, 2, and 3 are descriptive of the location of flow profiles in the corresponding regions. No H1 profile can actually be established since h_n is infinite for channels of horizontal bed. The A1 profile is impossible since the value of h_n is not real for flow in adverse channels. Of these 13 profiles, 12 are for gradually-varied flow and 1 (i.e., C2) for uniform flow. The general characteristics of these profiles are given in Table 1.2 and the examples of profiles are shown in Fig. 1.3. Figure 1.3a shows an example of H2 and H3 profiles in an horizontal channel; Fig. 1.3b shows M1, M2 and M3 profiles in a mild slope channel; Fig. 1.3c shows the critical profiles C1 and C3 for flow in a critical condition; Fig. 1.3d presents S1, S2 and S3 profiles in a steep channel; and Fig. 1.3e presents the A2 and A3 profiles in an adverse channel. It should be noted that a continuous flow profile usually occurs only in one region.

In addition, for the purpose of discussion, channel slopes may be classified as sustaining or non-sustaining, as mentioned in the book of Chow (1959). Sustaining slopes are those falling in the direction of flow, including mild, critical and steep slopes, and they are all positive slopes. A non-sustaining slope may be either horizontal or adverse. The bottom slope of a horizontal channel is zero. An adverse slope is a negative slope that rises in the direction of flow. The weight component of the flow acts along the slope in the flow direction in the case of sustaining slopes, while that is oppose to the flow direction in the case of non-sustaining slopes. In practice, only short lengths of non-sustaining slopes may be encountered.

Table 1.2 Types of GVF profiles

Channel slope category	Zone	Relation of h to h_n and h_c	Type of curves	Type of flow
1. Horizontal bed	2	$h > h_c$	H2:drawdown	Subcritical
($h_n = \infty$)	3	$h < h_c$	H3:backwater	Supercritical
2. Mild slope	1	$h > h_n > h_c$	M1:backwater	Subcritical
	2	$h_n > h > h_c$	M2:drawdown	Subcritical
	3	$h_n > h_c > h$	M3:backwater	Supercritical
3. Critical slope	1	$h > h_c$	C1:backwater	Subcritical
($h_n = h_c$)	3	$h < h_c$	C3:backwater	Supercritical
4. Steep slope	1	$h > h_c > h_n$	S1:backwater	Subcritical
	2	$h_c > h > h_n$	S2:drawdown	Supercritical
	3	$h_c > h_n > h$	S3:backwater	Supercritical
5. Adverse slope	2	$h > h_c$	A2:drawdown	Subcritical
	3	$h < h_c$	A3:backwater	Supercritical

1.6 Hydraulic Exponents

Almost all major hydraulic engineering activities in open channel flow involve the computation of the GVF profile. The computation of the GVF profile using direct integration method usually involves the determination of the hydraulic exponents, M and N for critical and uniform flows, respectively. Taking the natural logarithm of both sides of (1.17) and (1.19) and then differentiating them with respect to flow depth h, respectively, to obtain the following relations without the assumption that the coefficients C_z and C_k as well as on hydraulic exponents M and N are independent of the flow depth h:

$$\frac{2}{Z}\frac{dZ}{dh} = \frac{d \ln C_z}{dh} + \frac{dM}{dh} \ln h + \frac{M}{h} \tag{1.43}$$

and

$$\frac{2}{K}\frac{dK}{dh} = \frac{d \ln C_k}{dh} + \frac{dN}{dh} \ln h + \frac{N}{h}. \tag{1.44}$$

Fully neglecting the effect of h on C_z, C_k, M and N as done by Chow (1955), we are led to the following simplified relations

$$\frac{2}{Z}\frac{dZ}{dh} = \frac{M}{h} \tag{1.45}$$

and

$$\frac{2}{K}\frac{dK}{dh} = \frac{N}{h}. \tag{1.46}$$

Considering $Z^2 = A^3/T$ and $K^2 = A^2 R^{4/3}/n^2$ (if the Manning formula in SI units is used), the left hand sides of (1.41) and (1.43) can be expressed as followings

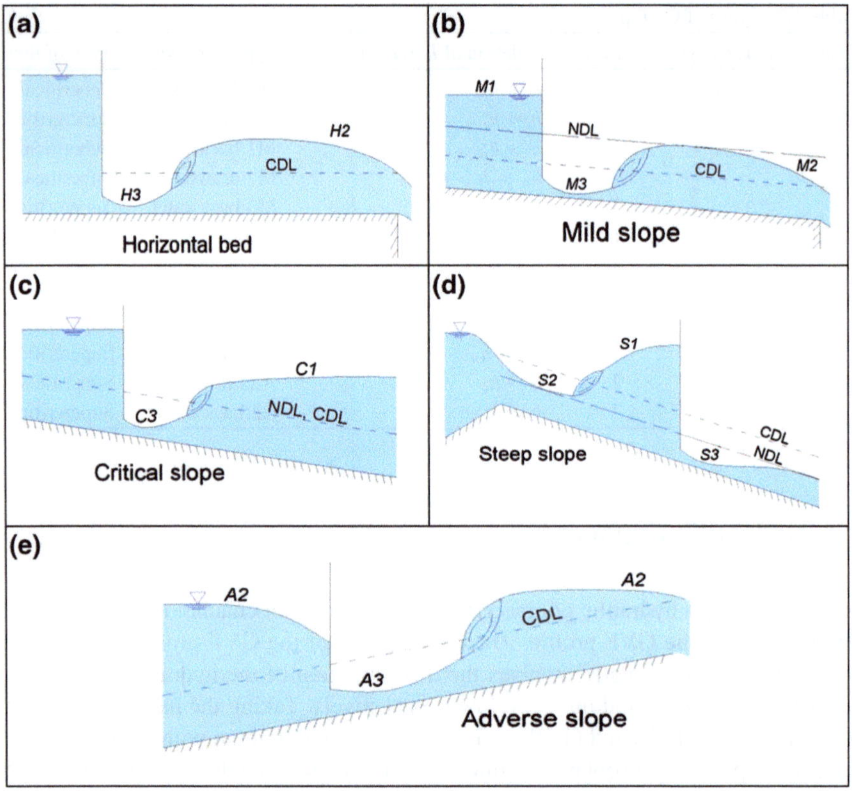

Fig. 1.3 Examples of profiles of gradually-varied flow in a long prismatic channel. **a** Shows an example of H2 and H3 profiles in an horizontal channel, **b** M1, M2 and M3 profiles in a mild slope channel, **c** the critical profiles C1 and C3 for flow in a critical condition, **d** S1, S2 and S3 profiles in a steep channel, and **e** the A2 and A3 profiles in an adverse channel

$$\frac{2}{Z}\frac{dZ}{dh} = \frac{1}{Z^2}\frac{dZ^2}{dh} = \frac{T}{A^3}\left(\frac{3A^2}{T}\frac{dA}{dh} - \frac{A^3}{T^2}\frac{dT}{dh}\right) = \frac{1}{A}\left(3T - \frac{A}{T}\frac{dT}{dh}\right) \tag{1.47}$$

and

$$\begin{aligned}\frac{2}{K}\frac{dK}{dh} = \frac{1}{K^2}\frac{dK^2}{dh} &= \frac{P^{4/3}}{A^{10/3}}\left(\frac{10}{3}\frac{A^{7/3}}{P^{4/3}}\frac{dA}{dh} - \frac{4}{3}\frac{A^{10/3}}{P^{7/3}}\frac{dP}{dh}\right)\\ &= \frac{2}{3}\left(\frac{5T}{A} - \frac{2}{P}\frac{dP}{dh}\right).\end{aligned} \tag{1.48}$$

Equating the right hand sides of (1.45) and (1.47), and equating the right hand sides of (1.46) and (1.48), we can obtain the expressions for M and N, respectively, as shown below.

$$M = \frac{h}{A}\left(3T - \frac{A}{T}\frac{dT}{dh}\right) \tag{1.49}$$

and

$$N = \frac{2h}{3}\left(\frac{5T}{A} - \frac{2}{P}\frac{dP}{dh}\right). \tag{1.50}$$

The above two equations show that both M and N are functions of channel section and the depth of flow. For a rectangular channel section having a bottom width B, the section area $A = Bh$, the wetted perimeter $P = B + 2h$, the hydraulic exponents M and N, according to (1.49) and (1.50), can be written as

$$M = \frac{h}{Bh}\left(3B - \frac{Bh}{B}\frac{dB}{dh}\right) = 3 \tag{1.51}$$

and

$$N = \frac{2h}{3}\left(\frac{5B}{Bh} - \frac{4}{B + 2h}\right) = \frac{10}{3} - \frac{8}{3}\frac{(h/B)}{1 + 2(h/B)}. \tag{1.52}$$

Equations (1.51) and (1.52) show that $M(= 3)$ is a constant but N depends on the inverse of aspect ratio (h/B) for a rectangular channel. It is obvious that for a wide channel, $h/B \to 0$, both M and N are constants, having values as $M = 3$ and $N = 10/3$. For narrow rectangular channels $(h/B \to \infty)$, $N \to 2$, according to (1.52). For a trapezoidal channel section having a bottom width B and side slopes 1 on z, the section area $A = h(B + zh)$, the wetted perimeter $P = B + 2h\sqrt{1 + z^2}$, M and N can be written as

$$M = \frac{3[1 + 2z(h/B)]^2 - 2z(h/B)[1 + z(h/B)]}{[1 + 2z(h/B)][1 + z(h/B)]} \tag{1.53}$$

and

$$N = \frac{10}{3}\frac{1 + 2z(h/B)}{1 + z(h/B)} - \frac{8}{3}\frac{\sqrt{1 + z^2}(h/B)}{1 + 2\sqrt{1 + z^2}(h/B)}. \tag{1.54}$$

According to (1.53), we can obtain that the values of M for rectangular channels (i.e., $z = 0$) and triangular channels (i.e., $h/B = \infty$, due to $B = 0$) are 3.0 and 5.0, respectively, while for trapezoidal channels in the range of 3.0–5.0, as shown in Fig. 1.4. According to (1.54), we can obtain that the values of N for wide channels (i.e., $h/B \to 0$) and triangular channels (i.e., $h/B = \infty$) are 10/3 and 16/3, respectively, if the Manning formula is used, while for trapezoidal channels N-values are in the range 2.0–5.33, as shown in Fig. 1.5.

For a circular channel section having a diameter d_0, according to (1.49) and (1.50), M and N can be written as

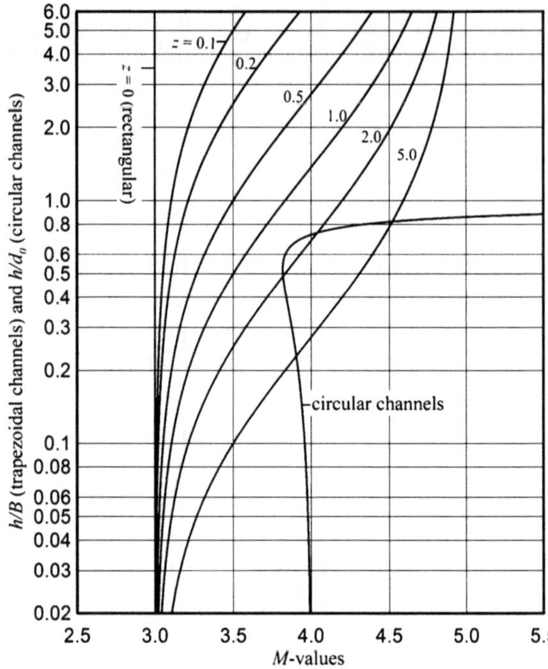

Fig. 1.4 Variations of the hydraulic exponent M on the dimensionless flow depth for trapezoidal and circular channels

$$M = \frac{h}{A}\left(3T - \frac{A}{T}\frac{dT}{dH}\right) = \frac{(h/d_0)}{(A/d_0^2)}\left[\frac{3T}{d_0} - \frac{2(A/d_0^2)}{(T/d_0)^2}\left(1 - \frac{2h}{d_0}\right)\right] \tag{1.55}$$

and

$$N = \frac{2h}{3}\left(\frac{5T}{A} - \frac{2}{P}\frac{dP}{dh}\right) = \frac{2(h/d_0)}{3(A/d_0^2)}\left(\frac{5T}{d_0} - \frac{4(R/d_0)}{(T/d_0)}\right), \tag{1.56}$$

in which the dimensionless section area (A/d_0^2), dimensionless wetted perimeter (T/d_0) and dimensionless hydraulic radius (R/d_0) for a circular channel are all dependent on the dimensionless water depth (h/d_0) as follows.

$$\frac{T}{d_0} = \sqrt{1 - \left(1 - \frac{2h}{d_0}\right)^2}, \tag{1.57}$$

$$\frac{A}{d_0^2} = \frac{1}{8}\left(2\arccos\left(1 - \frac{2h}{d_0}\right) - \sin\left[2\arccos\left(1 - \frac{2h}{d_0}\right)\right]\right), \tag{1.58}$$

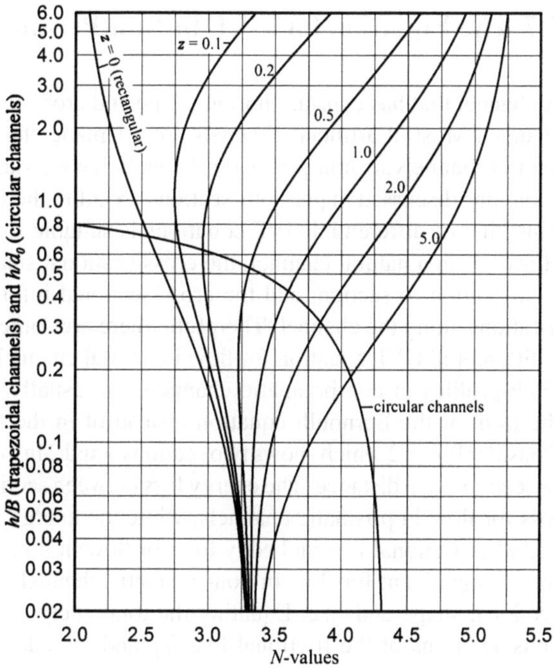

Fig. 1.5 Variation of the hydraulic exponent N on the dimensionless flow depth for trapezoidal and circular channels

$$\frac{R}{d_0} = \frac{1}{4}\left(1 - \frac{\sin\left[2\arccos\left[1 - \left(\frac{h}{d_0}\right)\right]\right]}{2\arccos\left[1 - 2\left(\frac{h}{d_0}\right)\right]}\right). \tag{1.59}$$

Since the change in depth of a gradually-varied flow is generally small, M and N may be assumed constant within the range of the limits of integration, as indicated by Chow (1955, 1959) by the classical direct integration method for the solution of the GVF equation. The method assumed constant values of M and N based on the averaged GVF depth in the concerned reach. The assumption of constant hydraulic exponents is satisfactory in the evaluation the GVF profiles in most rectangular and trapezoidal channels. However, the hydraulic exponents may vary appreciably with respective to the flow depth when the channel section has abrupt changes in geometry of cross section or is topped with a gradually closing crown, such as circular channels. In such cases, the channel length should be divided into a number of reaches in each of which the hydraulic exponents appear to be constant (Chow 1959), or an improved method or innovative method is needed to overcome this problem (Zaghloul 1998). Zaghloul and Shahin (1993) presented a software package to numerically solve the GVF profiles in circular channels by using the direct step method instead of the direct integration method. A review of attempts made by previous investigators on the hydraulic exponents will be presented in Sect. 2.6 of Chap. 2.

1.7 The Equation for GVF in Non-Prismatic Channels

A channel that has constant bottom slope and cross-section is termed as a prismatic channel. Most of artificial channels are prismatic channels over long stretches. All natural channels are non-prismatic channels, having varying cross-sections. The GVF equations discussed in previous sections are all in the form of differential equations. This kind of differential GVF equations is suitable for GVF in prismatic channels. However, in a natural channel, the cross-sectional shape and size are likely to vary from section to section, and the cross-sectional information is known only a few locations along the channel. Therefore, there are some basic difficulties to establish a differential GVF equation for flow in natural channels. For the computation of the GVF profiles in non-prismatic channels, we usually solve the energy equation in the form of the Bernoulli equation instead of in the form of differential equation. Consider Fig. 1.2 which shows two sections 1 and 2 in a channel. Section 1 is upstream of section 2 at a distance. The energy loss between section 1 and 2 is mainly frictional loss for flow in prismatic channels, while the energy loss between section 1 and 2 includes frictional loss and eddy loss for flow in non-prismatic channels. The eddy loss is significant for flow in non-prismatic channels due to the variation of cross-sectional shape and size. Equating the total energies at sections 1 and 2, with the considerations of the frictional loss h_f and the eddy loss h_e, yields a basic GVF equation for flow in a non-prismatic channel as shown in (1.60), if calculation is assumed to proceed downstream.

$$z_{b1} + y_1 + \frac{\alpha_1 V_1^2}{2g} = z_{b2} + y_2 + \frac{\alpha_2 V_2^2}{2g} + h_f + h_e \tag{1.60}$$

The frictional loss can be estimated by using Manning formula. There is no rational method for estimating the eddy loss. The eddy loss is usually assumed to be proportional to the velocity-head difference between two adjacent sections. An alternative practice of accounting for eddy losses is to increase the Mannings roughness coefficient by a suitable small amount. This procedure simplifies calculations in some cases, as pointed out by Subramanya (2009). The problem of computation of the GVF profile in a non-prismatic channel can be stated as: Given the discharge and stage at one section and the cross-sectional information at two adjacent sections, it is required to determine the stage at the other section. The sequential determination of the stage as a solution of the assigned problem will lead to the GVF profile, as mentioned by Chow (1959) and Subramanya (2009). Nowadays most GVF profiles are numerically computed by using commercial programs, such as HEC-RAS computer program which is a public domain code developed by the U. S. Army Corps of Engineers (USACE 2002). The numerical computation of GVF profiles in non-prismatic channels is out of the scope of this book.

1.8 Summary

Many hydraulic engineering activities involve the computation of the water surface profiles of gradually-varied flow (GVF) in open channels. This chapter has reviewed and discussed the derivation of one-dimensional GVF equation based on the energy conservation law. The hydraulic exponents *M* and *N* for the critical-flow computation and uniform-flow computation are introduced in the derivation of the GVF equation. Equation (1.41) is the final form of the GVF equations for flow in sustaining channels. This equation describes the relation of the slope of GVF profile and its corresponding flow depth with the critical depth, normal depth and two hydraulic exponents as parameters. Based on the GVF equation, the classification of GVF profiles is also introduced in this chapter. The final form of the GVF equation for flow in adverse channels (having a negative channel slope) is presented in (1.42) with a fictitious normal depth. Since the change in depth of a gradually varied flow is generally small, the hydraulic exponents may be assumed constant within the range of the limits of integration in the classical direct integration method for the solutions of the GVF equation. The assumption of constant hydraulic exponents is satisfactory in the evaluation the GVF profiles in most rectangular and trapezoidal channels. However, the hydraulic exponents may vary appreciably with respective to the flow depth when the channel section has abrupt changes in geometry of cross section or is topped with a gradually closing crown, such as circular channels. An innovative method to get analytical solution of the GVF equation by using the direct integration method and the Gaussian hypergeometric functions will be introduced in the next chapters. The analytical solutions of GVF profiles so obtained in this book are basically suitable for flow in prismatic channels.

References

Chaudhry M (2006) Open-channel flow. Springer, New York

Chen CL (1991) Unified theory on power laws for flow resistance. J Hydraul Eng ASCE 117(3):371–389

Chen CL (1992) Momentum and energy coefficients based on power-law velocity profile. J Hydraul Eng ASCE 118(11):1571–1584

Chen CL (1995) Free-surface instability criterion as affected by velocity distribution. J Hydraul Eng ASCE 121(10):736–743

Chow VT (1981) Hydraulic exponents. J Hydraul Div ASCE 107(HY11):1489–1499

Chow VT (1955) Integrating the equation of gradually varied flow. Proceedings Paper No. 838. ASCE 81:1–32

Chow VT (1959) Open-channel hydraulics. McGraw-Hill, New York

Jan CD, Chen CL (2012) Use of the Gaussian hypergeometric function to solve the equation of gradually varied flow. J Hydrol 456–457:139–145

Kranenburg C (1990) On the stability of gradually varying flow in wide open channel. J Hydraul Res 28(5):621–628

Subramanya K (2009) Flow in open channels, 3rd edn. McGraw-Hill, Singapore

USACE (2002) HEC-RAS, River analysis system, vol 3.1. U. S. Army Crops of Engineers Hydrologic Engineering Center. CPD-68

Vatankhah AR (2012) Direction integration of Manning-based GVF equation in trapezodial channels. J Hydrol Eng ASCE 17(3):455–462

Zaghloul NA (1998) Hydraulic exponents m and n for gravity flow pipes. Adv Water Resour 21(3):185–191

Zaghloul NA, Shahin MM (1993) Computer simulation of gradually varied profiles in circular sections. Adv Eng Softw 16:37–46

Chapter 2
Conventional Integral Solutions of the GVF Equation

2.1 Introduction

The gradually-varied flow (GVF) equation for flow in open channels is normalized using the normal depth, h_n, before it can be analytically solved by the direct integration method. The right-hand side of the dimensionless GVF equation is a rational function expressed in terms of the dimensionless flow depth, h/h_n, to the two powers, M and N (called the hydraulic exponents), with the ratio of the critical depth, h_c, to h_n (or h_c/h_n) as a parameter. For simplicity, M and N are usually treated as constant, thus averting the impasse in integrating directly the GVF equation with the h/h_n-varying M and N. The direct-integration method is a conventional method used to analytically solve the GVF equation. The varied-flow function (VFF) needed in the direct-integration method has a drawback caused by the imprecise interpolation of the VFF-values. To overcome the drawback, we use the Gaussian hypergeometric functions (GHF) to analytically solve the GVF equation without recourse to the VFF. This chapter presents a review of conventional integral solutions of the GVF profiles by using the varied-flow function (VFF) and/or the elementary transcendental functions (ETF).

2.2 GVF Solution in Terms of Varied-Flow Function

The gradually varied flow (GVF) is a steady non-uniform flow in a prismatic channel with gradual changes in its water surface elevation. The computation of GVF profiles involves basically the solution of the dynamic equation of GVF. The main objective of the computation is to determine the shape of the flow profile. As shown by (1.41) in Chap. 1, the one-dimensional GVF dynamic equation for flow in open channels with arbitrary cross-sectional shapes can be expressed in terms only of the flow depth, h, if the section factor, Z, and the conveyance, K, of the channel section can be expressed in terms of the power of h with respective hydraulic exponents, M and N,

C.-D. Jan, *Gradually-varied Flow Profiles in Open Channels*,
Advances in Geophysical and Environmental Mechanics and Mathematics,
DOI: 10.1007/978-3-642-35242-3_2, © Springer-Verlag Berlin Heidelberg 2014

as shown in (1.17) to (1.20). For the convenience of discussion, we re-write (1.41) for small slope angle in the form as

$$\frac{dh}{dx} = S_0 \frac{1 - (h_n/h)^N}{1 - (h_c/h)^M}. \tag{2.1}$$

Normalizing the flow depth h by the normal depth h_n, namely $u = h/h_n$, and substituting the dimensionless depth u into (2.1) yields

$$\frac{du}{dx} = \frac{S_0}{h_n} \frac{1 - u^{-N}}{1 - (h_c/h_n)^M u^{-M}}. \tag{2.2}$$

Even though it is commonly used to represent the velocity, the symbol u has been conventionally used to represent the dimensionless flow depth in the analysis of GVF profiles, as shown in some text books, such as Chow (1959), Subramanya (2009), among others. Hence, we will use the symbol u to represent the dimensionless flow depth in this book.Rearranging the equation into its reciprocal form, we can get an expression for dx in the form as

$$dx = \frac{h_n}{S_0} \left[1 - \frac{1}{1 - u^N} + \left(\frac{h_c}{h_n} \right)^M \frac{u^{N-M}}{1 - u^N} \right] du. \tag{2.3}$$

Given a cross-sectional shape other than wide rectangle, the M- and N-values generally vary with h/h_n; therefore, we cannot integrate the h_n-based dimensionless GVF equation over h/h_n. As long as the change in depth of a gradually varied flow is generally small, for simplicity, the hydraulic exponents are traditional assumed constant within the range of the limits of integration as was done by Chow (1959) and others. Integrating (2.3) yields

$$x = \frac{h_n}{S_0} \left[u - \int_0^u \frac{1}{1 - u^N} du + \left(\frac{h_c}{h_n} \right)^M \int_0^u \frac{u^{N-M}}{1 - u^N} du \right] + \text{Const.}. \tag{2.4}$$

The GVF equation so formulated as shown in (2.1) or (2.2) is a nonlinear differential equation of first order in h with the two exponents, M and N, and the ratio of the critical depth, h_c, to the normal depth, h_n, as three parameters. Even though Z and K can be expressed in terms of the power of h with the respective exponents, M and N, as shown in Chap. 1, one has not yet successfully integrated directly such a nonlinear differential equation with arbitrarily assumed real numbers of M and N due to the difficulty in integrating the two integrals of a proper fraction, which are expressed as reciprocals of the rational function representing the slope of the GVF profile, as shown in (2.4).

As shown in the text book of Chow (1959), the first integral on the right hand side of (2.4) is designated by $F(u, N)$ that is known as the varied-flow function (VFF), namely

$$F(u, N) = \int_0^u \frac{1}{1 - u^N} du. \tag{2.5}$$

The second integral of the right hand side of (2.4) may also be expressed in the form of VFF after a variable transformation of u. Letting $v = u^{N/J}$ and $J = N/(N - M + 1)$, we can express the second integral in the form as (Chow 1959)

$$\int_0^u \frac{u^{N-M}}{1 - u^N} du = \frac{J}{N} \int_0^v \frac{1}{1 - v^J} dv = \frac{J}{N} F(v, J), \tag{2.6}$$

where $F(v, J)$ is a varied flow function like $F(u, N)$, except the variable u and the exponent N are replaced by v and J, respectively. After (2.5) and (2.6) are substituted in (2.4), the solution of length along the channel for GVF profile can be expressed in terms of these two varied-flow functions, as shown below.

$$x = \frac{h_n}{S_0} \left[u - F(u, N) + \left(\frac{h_c}{h_n} \right)^M \frac{J}{N} F(v, J) \right] + \text{Const.}. \tag{2.7}$$

Equation (2.7) contains two VFF, and its solution can be simplified by the use of the VFF tables as given in the book of Chow (1959). It was said that the preparation of the VFF table was undertaken and performed for the first time during 1914–1915 by the Research Board of the then Russian Reclamation Service under the direction of Boris A. Bakhmeteff. The VFF table was published in 1932 when Bakhmeteff became Professor of Civil Engineering at Columbia University. More than 20 years later, the VFF table was extended and improved by Ven Te Chow during 1952–1954 and published in 1955. Chow's VFF table for positive and negative slopes is a big table having 15 pages as shown in the Appendix D in the book of Chow (1959). The literature review about the use of VFF in the solution of GVF profile will be shown in the following sections.

2.3 GVF Solution by the Bresse Method

For a wide rectangular channel, if the Chézy formula is used, the hydraulic exponents $M = N = 3$. Thus $J = N = 3$, $v = u$, and (2.7) becomes a simpler form.

$$x = \frac{h_n}{S_0} \left[u - \left(1 - \left(\frac{h_c}{h_n} \right)^3 \right) F(u, 3) \right] + \text{Const.}. \tag{2.8}$$

The function $F(u, 3)$ was first evaluated by Bresse (1860) in a closed form as

Table 2.1 The derivation of the Bresse solution of the varied-flow function $F(u, 3)$ as shown in Eq. (2.9)

Integral function: $F(u,3) := \int_0^u \frac{1}{1-u^3} du$
solution: $\int_0^u \frac{1}{1-u^3} du = \int_0^u \frac{1}{(1-u)(u^2+u+1)} du = \frac{1}{3}\int_0^u \frac{1}{1-u} du + \frac{1}{3}\int_0^u \frac{u+2}{(u^2+u+1)} du = \frac{1}{3}\int_0^u \frac{1}{1-u} du + \frac{1}{6}\int_0^u \frac{2u+1}{(u^2+u)+1} du + \frac{1}{2}\int_0^u \frac{1}{(u^2+u+1)} du = -\frac{1}{3}\ln(1-u) + \frac{1}{6}\ln\left(u^2+u+1\right) + \frac{1}{\sqrt{3}}\arctan\left(\frac{2u+1}{\sqrt{3}}\right) - \frac{1}{\sqrt{3}}\arctan\left(\frac{1}{\sqrt{3}}\right) = \frac{1}{6}\ln\left(\frac{u^2+u+1}{(1-u)^2}\right) + \frac{1}{\sqrt{3}}\arctan\left(\frac{2u+1}{\sqrt{3}}\right) - \frac{1}{\sqrt{3}}\arctan\left(\frac{1}{\sqrt{3}}\right)$

$$
\begin{aligned}
F(u,3) &= \int_0^u \frac{1}{1-u^3} du \\
&= \frac{1}{6}\ln\left(\frac{u^2+u+1}{(u-1)^2}\right) + \frac{1}{\sqrt{3}}\arctan\left(\frac{2u+1}{\sqrt{3}}\right) - \frac{1}{\sqrt{3}}\arctan\left(\frac{1}{\sqrt{3}}\right),
\end{aligned}
\tag{2.9}
$$

or in another form as

$$
\begin{aligned}
F(u,3) &= \int_0^u \frac{1}{1-u^3} du \qquad (2.10) \\
&= \frac{1}{6}\ln\left(\frac{u^2+u+1}{(u-1)^2}\right) - \frac{1}{\sqrt{3}}\operatorname{arccot}\left(\frac{2u+1}{\sqrt{3}}\right) + \frac{1}{\sqrt{3}}\operatorname{arccot}\left(\frac{1}{\sqrt{3}}\right),
\end{aligned}
$$

Since there is a relation between two inverse trigonometric functions, i.e.,

$$
\arctan\left(\frac{2u+1}{\sqrt{3}}\right) + \operatorname{arccot}\left(\frac{2u+1}{\sqrt{3}}\right) = \frac{\pi}{2}. \tag{2.11}
$$

The details of the derivation of (2.9) are shown in Table 2.1. A determination of GVF profile by using (2.8) and (2.9) [or (2.10)] is widely known as the Bresse method (Chow 1959). Obviously, the GVF solution by the Bresse method is only limited for flow in a wide rectangular channel and its flow resistance can be described by the Chézy formula. It will be discussed later that the Bresse solution is one of the solution using elementary transcendental functions.

2.4 GVF Solution by the Bakhmeteff-Chow Procedure

In general cases, the hydraulic exponents are some positive real numbers, depending channel geometry and flow depth, and this causes the difficuly in the integration of two integrals in (2.4). To overcome such difficulty in the integration, Bakhmeteff (1932) prepared an integration table of the varied-flow function (VFF), by which such two integrals for the fixed M- and N-values could be approximately evaluated. Decades later, Chow (1955) refined and extended the VFF table for computing GVF profiles numerically in both sustaining and adverse channels with all kinds of regular cross-sectional shapes. The solution of the length x of the flow profile can be simplified by the use of the table of VFF (Chow 1959).

Hydraulic engineers resorted to a procedure proposed by Bakhmeteff (1932) and Chow (1959), who evaluated the two integrals using the VFF table, for more than a half century for there has been no method better than the Bakhmeteff-Chow procedure based on the VFF table to evaluate such two integrals, from which GVF profiles are computed numerically for channels with h-dependent M- and N-values. Thus, if we want to evaluate these two integrals more precisely by some means other than the VFF table, we should first examine how Bakhmeteff and Chow computed the VFF-values, which enabled them to construct the VFF table. We undertake this scrutiny next.

In the approach taken by Bakhmeteff (1932), because his GVF equation is equivalent to the one for GVF in wide channels with the flow resistance expressed by the Chézy formula, namely $M = N = 3$, the second integral can merge into the first integral (similiar to 2.8) to form a sole integral, which he called the VFF. Four methods were proposed by him in the computation of the VFF-values, as explained in Appendix II of his book. In all the four methods except for the third one in which he used the approximate integration formula, he expanded the integrand of the VFF into an infinite series, thereby numerically integrating and calculating a sum of the first few terms of the expanded infinite series for each fixed value of N. The variable and the parameter of the VFF are the h_n-based dimensionless depth, h/h_n, and N, respectively. He determined a number of terms needed in the computation of the infinite series for each method by a specific convergence criterion established for each infinite series. His fourth method was particularly designed to safeguard the convergence of the computed VFF-value for h/h_n near unity in case that his third method failed to yield the satisfactory result. As claimed by Bakhmeteff, the precision of the computed VFF-values in his table was to keep their numerical errors less than 0.0005. It was also reported that the VFF-values so computed for the specific N-values in the range between 2.8 and 5.4 were tabulated by a team of Russian scientists led by Bakhmeteff before the Russian revolution in 1917, while the range of the N-values in the VFF table, especially its lower end, was slightly extended after the Russian revolution.

For integrating the GVF equation, Chow (1955) essentially used Bakhmeteff (1932) four methods in the computation of the VFF-values for the first integral. The integrand in the second integral was first transposed to a form of the VFF of a composite variable with a modified parameter expressed in terms of M and N before it was evaluated using the same four methods to compute the VFF-values in the evaluation of the first integral, as shown in (2.6). After recalculating the VFF-values for the two integrals, Chow (1955) corrected several errors found in Bakhmeteff's (1932) VFF table for the N-values ranging from 2.8 to 5.4 and then extended them to cover the N-values ranging from 2.2 to 9.8 for sustaining slopes. Besides, Chow (1957) prepared the VFF table for adverse slopes for the N-values in the range between 2.0 and 5.5. For dissemination of the Bakhmeteff-Chow procedure based on such two VFF tables, one for sustaining slopes and the other for adverse slopes, Chow (1959) further published both refined VFF tables in his widely-circulated book on open-channel hydraulics.

2.5 Drawbacks on the VFF Table for GVF Solution

Unfortunately, the Bakhmeteff-Chow procedure based on the VFF table suffers from two major drawbacks, which have impeded the progress of the 1-D approach in the GVF profile computation. The first drawback is caused by the imprecise interpolation of the VFF-values for a range of the dimensionless flow depth near unity or between the contiguous N-values as the VFF parameter. In other words, the VFF-values provided in the table are not sufficiently precise to evaluate such two integrals, especially for the dimensionless flow depth near unity, where the VFF-value changes rapidly with a small change in the dimensionless flow depth or between the contiguous N-values as the VFF parameter. On the other hand, the second drawback has resulted from the incompleteness of a method proposed by Bakhmeteff (1932); Chow (1955, 1959) to render the exact M- and N-values. Rigorously speaking, to use their method, the computed M- and N-values, which are supposed to vary considerably with the depth of flow in a channel with cross-sectional shape other than wide rectangle, are inaccurate. In fact, to overcome the first drawback, we have undertaken a novel approach to integrate the two integrals using the Gaussian hypergeometric function (GHF) without recourse to the VFF table as shown in Jan and Chen (2012) as well as in Jan and Chen (2013). The elaboration of this technique using the h_n-based GHF is reported in Chap. 3 and that using the h_c-based GHF in Chap. 4. However, before presenting such a technique, we review briefly all worthwhile efforts made by previous investigators to overcome both drawbacks in the following.

2.6 Attempts Made on M and N by Previous Investigators

In the past, attention has been focused on how to overcome the second drawback, but all attempts made by previous researchers to cope with it have not produced very fruitful results. Though overcoming the second drawback is not the main objective of the study of this book, it deserves a cursory review because knowledge of our failure to fix the second drawback may help counter the first drawback. Mononobe (1938) introduced two assumptions for the M- and N-values, thereby taking into account the possible effects of changes in velocity, friction head, and channel shape without dividing the channel length into short reaches. To compute the M- and N-values, various methods similar to that proposed by Bakhmeteff (1932); Chow (1955, 1959) or slight modifications thereof were adopted by many investigators, such as Woodward and Posey (1941), Kirpich (1948), Lee (1947), Von Seggern (1950), Subramanya and Ramamurthy (1974), Patil et al. (2001, 2003), Zaghloul and Anwar (1991), and Srivastava (2003), among others.

All the methods used by the previous investigators assume constant M and N. To remove the constant assumption of M and N, Chow (1981) formulated the two linear differential equations of first order, one for M and the other for N, thereby solving them for the exact expressions of the M- and N-values, which vary with the flow depth, h. Though Chow (1981) succeeded in removing the unwanted assumption of constant M and N from his earlier method, one can readily prove that the formulation of the two linear differential equations of first order for M and N are redundant, for the solutions of such differential equations are indeed identical to the M- and N-expressions that can be derived from the respective definitions of M and N. Chow (1981), aside from a lack of the theoretical basis to justify the derivation of his linear differential equations, could not even analytically solve the equations for h-dependent M and N for channels with any regular cross-sectional shape except for triangle and wide rectangles. In practical applications of Chow (1981) method, for example, to partially full flows in closed conduits, such as circular pipes, where the M- and N-values vary rapidly as the flow depth approaches the crown, Zaghloul (1998) numerically solved linear differential equations numerically for M and N proposed by Chow (1981) using the Runge–Kutta method. Although Zaghloul (1998) applied Chow's (1981) method to express the h-dependent M- and N-values for gravity flow in pipes, Shivareddy and Viswanadh (2008) a decade later still did not take into account the respective variations of M and N with h for flow in channels with D-shaped tunnel section.

2.7 Previous Studies on Integrating the GVF Equation

As for the first drawback, one cannot completely fix it until a new technique is developed to evaluate the two integrals of the proper fraction without resort to the VFF table. Apparently, the difficulty in overcoming the first drawback has resulted from a lack of our mathematical knowledge to find the exact functions to integrate such two integrals, which are transposed from the GVF equation, i.e. the nonlinear differential equation of first order in h with M, N, and h_c/h_n as the three parameters. Because the rational function representing the reciprocal of the slope of the GVF profile is improper (i.e., the degree of the denominator is less than or equal to the degree of the numerator), it can be rewritten by the process of long division (i.e., division algorithm) as a polynomial plus a proper fraction. A literature survey reveals that many studies have been undertaken by previous investigators to integrate such a proper fraction. For expedience, this proper fraction can be separated into two component fractions and the integration of the two component fractions has hitherto been referred to as the two integrals. The integration of both integrals can be performed in two ways. The first way is to have each integrand in the two integrals expanded into a finite set of partial fractions so that every term of them can be integrated separately, using the elementary transcendental functions (ETF), such as trigonometric, inverse trigonometric, hyperbolic, inverse hyperbolic, exponential, and logarithmic functions. In the second way, each of the two integrals was integrated using a number of various infinite series, one in a specified range of the dimensionless depth, thereby computing the values of both integrals. Obviously, the four methods used by Bakhmeteff (1932); Chow (1955, 1957, 1959) in the computation of the VFF-values belong to the second way, and other typical studies attempted by previous investigators are reviewed chronologically in the following.

The Bresse solution of GVF profile as shown in (2.8) and (2.9) [or (2.10)], was probably the earliest ETF-based solution for $M = N = 3$ that one can find in the literature. In fact, the reciprocal of the dimensionless GVF equation for $M = N = 3$, which can be exactly integrated to yield the Bresse solution, was used by Bakhmeteff (1932) to formulate the two integrals. Such two integrals for $M = N = 3$ can be transposed and merged into a sole integral, which Bakhmeteff (1932) called the VFF. Later, in an analogous approach, Gibson (1934) obtained the same sole integral, which he called the backwater function instead of the VFF and then tabulated it for computing GVF profiles in channels with cross-sectional shape other than wide rectangle. Gunder (1943) was among the earliest persons who applied the method of partial-fraction expansion in the evaluation of the two integrals for $M = 3$ and $N = 10/3$. The two integrals upon expansion in partial fractions were integrated by him using the ETF. In fact, the two integrals so evaluated appears to be a sum of those terms expressed later in Equations (D.2) and (D.6) as shown in Appendix D for flow in wide channels with flow resistance expressed using the Manning formula.

The two integrals formulated by Von Seggern (1950) look slightly different from those derived by Bakhmeteff (1932); Chow (1955, 1959), but in fact the integrands of both integrals prove to be identical if the former is converted to the latter, or vice versa, because he adopted the reciprocal of the dimensionless depth used by Bakhmeteff and Chow. Von Seggern also used an infinite series to compute the values of the two integrals. The infinite series used by Von Seggern is similar to that adopted by Bakhmeteff except with a slight modification made to account for the effect of the parameter, M, contained in one of the integrals. To meet the conversion criterion imposed on the integration of the infinite series, Von Seggern transposed the integrands of the two integrals into several plausible forms so that the values of both integrals at various ranges of his dimensionless depth can be approximately computed without inducing many terms needed in the numerical computation to warrant the convergence of the infinite series.

Instead of evaluating the two integrals, Pickard (1963) transposed the GVF equation into a form suitable for integration, i.e., the twelve flow profiles (three in mild, two in critical, three in steep, two in horizontal, and two in adverse channels) being arranged into four different integrals of two terms each. Any of the integrals was simply expressed in terms of one of three higher transcendental functions, which Pickard called the backwater integrals of the first, second, and third kinds. Among the three kinds, the backwater integral of the second kind was integrated in closed form, but the backwater of the first and third kinds were more conveniently expanded into an infinite series. The latter on integration yielded a solution with sort of the GHF before being reduced to a finite series of polylogarithms and polynomials of moderate order.

To integrate directly the GVF equation, most previous investigators preferred to express the rational function representing the reciprocal of the slope of the GVF profile in terms of the dimensionless flow depth rather than to substitute the geometric elements of a given channel section into the section factor, Z, and the conveyance, K, of the channel section, thereby evaluating the two integrals, but some investigators did it in reverse by substituting their geometric elements of a channel section under study into Z and K in the GVF equation before evaluating the two integrals. For example, Keifer and Chu (1955) computed backwater curves in closed conduits by substituting the geometric elements of a circular conduit section into Z and K in the GVF equation before integrating it by use of Simpsons rule. In another example, Allen and Enever (1968) substituted the geometric elements of a rectangular, triangular, or trapezoidal section into Z and K in the GVF equation before evaluating the two integrals. Allen and Enever used two power-law flow resistance formulas, the Chézy formula and the Blasius formula in their formulation of the GVF equation. They subsequently evaluated the two indefinite integrals via the partial-fraction expansion and then attempted to get the ETF-based solution for each of the cross-sectional geometries under study. Because a generalized form of the two integrals formulated using the 1/7th-power flow resistance formula for such regular channel shapes was too complex to be integrated directly, they only obtained a few simple ETF-based solutions, such as the Bresse solution for wide rectangular channels and a solution based on logarithmic function for narrow rectangular channels, among others.

In another approach analogous to the procedure used by Chow (1955, 1957, 1959) in the computation of the VFF, Gill (1976). evaluated the two indefinite integrals in two steps. Gill directly integrated the first integral using the partial-fraction expansion for 6 different N-values, i.e., $N = 2.5, 3, 10/3, 3.5, 4$ and 5 for each case of sustaining and adverse slopes, thereby obtaining the ETF-based solution of the first integral for each N-value studied. Upon acquiring all the ETF-based solutions of the first integrals for such N-values studied, Gill proceeded to evaluate the second integral by transposition to a form of the VFF of a composite variable with a modified parameter expressed in terms of M and N, by which he was able to express the second integral in the form of the first integral.

In an approach similar to that taken by Allen and Enever (1968), Kuma (1978) substituted their geometric elements of a rectangular or triangular channel section into Z and K in the GVF equation before evaluating the two integrals. However, unlike the two flow resistance formulas adopted by Allen and Enever for comparison, the Chézy formula was the only one used by Kumar in the GVF equation. Subsequently, Kumar also evaluated the two integrals via the partial-fraction expansion and then obtained their ETF-based solutions, thereby constructing two tables of his exclusively defined VFF, one for rectangular channels and the other for triangular channels, which are similar in concept to, but not refined as, the VFF tables developed by Bakhmeteff (1932); Chow (1955, 1957, 1959).

An approach taken by Zaghloul (1990, 1992) to integrate the GVF equation was the same as that used by Keifer and Chu (1955) to compute GVF profiles in circular channels by substituting the geometric elements of a circular conduit section into Z and K in the GVF equation before integrating it by use of Simpson's rule or the direct step and integration methods. Zaghloul formulated a computer model in such a way to handle all GVF computations in sustaining, horizontal, and adverse pipe slopes.

Instead of expressing the rational function representing the reciprocal of the slope of the GVF profile in the form of a polynomial plus a proper fraction, which was further separated into the two integrals, Ramamurthy et al. (2000) transformed (2.1) to (2.12) and then integrated directly.

$$S_0 \frac{dx}{dh} = \frac{1-(h_c/h)^M}{1-(h_n/h)^N}. \tag{2.12}$$

The right-hand side of (2.12) was expressed as a series in h_n/h, using the binomial series expansion to faciliate its direct integration. That is to say, the rational function in (2.12) was expanded in the form of a binomial series, i.e., an infinite series similar to that expanded by Bakhmeteff (1932) in his first method to compute the VFF. This binomial series merits attention because the infinite number of terms in the binomial series can be rearranged into two groups, each of which may be expressed in the form of an infinite series resembling the one that is defined using the GHF, as shown in Appendix A. Because the binomial series so expanded is valid only if the dimensionless flow depth or its reciprocal is less than unity, the binomial series expansion was applied by Ramamurthy et al. only in two zones of the GVF profiles in mild (M) channels corresponding to the M1 and M2 profiles. The CVF solutions obtained by Ramamurthy et al. are summarized as follows.

For $h_n/h < 1$, the expansion of (2.12) in the form as

$$\begin{aligned} S_0 \frac{dx}{dh} &= \frac{1-(h_c/h)^M}{1-(h_n/h)^N} \\ &= \left[1-\left(\frac{h_c}{h}\right)^M\right]\left[1+\left(\frac{h_n}{h}\right)^N+\left(\frac{h_n}{h}\right)^{2N}+\ldots\right]. \end{aligned} \tag{2.13}$$

The direct integration of (2.13) yields

$$x = \frac{h}{S_0}\left[\sum_{i=0}^{\infty}\frac{(h_n/h)^{iN}}{1-iN} - \left(\frac{h_c}{h}\right)^M \sum_{i=0}^{\infty}\frac{(h_n/h)^{iN}}{1-M-iN}\right] + \text{Const.}. \tag{2.14}$$

For $h_n/h > 1$, the expansion of (2.12) in the form as

$$S_0\frac{dx}{dh} = \frac{1-(h_c/h)^M}{1-(h_n/h)^N}$$
$$= \left(\frac{h}{h_n}\right)^N\left[\left(\frac{h_c}{h}\right)^M - 1\right]\left[1+\left(\frac{h}{h_n}\right)^N+\left(\frac{h}{h_n}\right)^{2N}+...\right]. \quad (2.15)$$

The direct integration of (2.15) yields

$$x = \frac{h}{S_0}\left[-\sum_{i=1}^{\infty}\frac{(h/h_n)^{iN}}{1+iN}+\left(\frac{h_c}{h}\right)^M\sum_{i=1}^{\infty}\frac{(h/h_n)^{iN}}{1-M+iN}\right]+\text{Const.}. \quad (2.16)$$

As reviewed briefly above, there have been two methods used by previous investigators to evaluate the two integrals. To assess the practical usefulness of the two methods to evaluate the two integrals, we compare the advantages and disadvantage of two solutions obtained from both methods in hopes of keeping abreast of the status quo, progress, and perspective of such evaluations as well as of finding some novel techniques to bring such evaluations to a justifiable end. As far as the accuracy of the computation is concerned, the ETF-based solution is superior to the infinite-series-based solution because the former is exact in contrast to the latter, which is only approximate pending the convergence rule to decide how many terms are required in the infinite series so as to be computed within the prescribed tolerance. On the other hand, if we assess the usefulness of both solutions from the practical point of view, the infinite-series-based solution should be more useful than the ETF-based solution because one can formulate a table, such as the VFF table, using data obtained from the infinite-series-based solution for a specified range of various N-values. Such a VFF-like table is useful in a numerical procedure for computing GVF profiles in channels with cross-sectional shape other than wide rectangle, whose M- and N-values vary with h. Unfortunately, it is too tedious to construct such a VFF-like table from the ETF-based solutions, though Gill (1976) obtained a number of ETF-based solutions for such purposes.

Nowadays, the extensive and rapidly growing use of digital computers can facilitate the evaluation of the two integrals, one way or the other, by use of the mathematics software, such as the Mathematica software (Wolfram 1996). As for the first way, for example, Venutelli (2004) found from the Mathematica software the ETF-based solutions of the two indefinite integrals for $M = 3$ and $N = 10/3$ (provided the Manning formula is used as flow resistance), which are expanded into two finite sets of partial fractions, each term of which can be integrated separately by use of the ETF. Equation (2.4) on substitution of $M = 3$ and $N = 10/3$ yields

$$x = \frac{h_n}{S_0}\left[u - \int_0^u \frac{1}{1-u^{10/3}}du + \left(\frac{h_c}{h_n}\right)^3\int_0^u \frac{u^{1/3}}{1-u^{10/3}}du\right]+\text{Const.}. \quad (2.17)$$

Venutelli did not directly solve (2.17), but made a substitution of $\eta = u^{1/3}$ on the two integral terms of (2.17) first so as to transform the equation in the following form.

$$x = \frac{h_n}{S_0}\left[u - \int_0^{\eta} \frac{3\eta^2}{1-\eta^{10}}d\eta + \left(\frac{h_c}{h_n}\right)^3 \int_0^{\eta} \frac{3\eta^3}{1-\eta^{10}}d\eta\right] + \text{Const..} \qquad (2.18)$$

Venutelli then solved the two integral terms of (2.18) by using the Mathematica software so as to obtain an ETF-based solution of GVF profiles for subcritical and supercritical flows in mild and steep wide channels. In fact, the transformation from (2.17) to (2.18) is not necessary. Nowadays, to get an ETF-based solution of the dimensionless GVF profile for flow in a wide channel with hydraulic exponents $M = 3$ and $N = 10/3$, we can directly from (2.17) by using the Mathematica software, as shown in Table 2.2. More ETF-based solutions of the two indefinite integrals having $M = 3$ and five different N-values obtained from the Mathematica software are displayed in Appendix D. It is obvious that these ETF-based analytical solutions, as shown in Appendix D, are lengthy and disorgannized, except Bresse's solution for the case of $M = N = 3$.

As for the second way to express the solutions of the two integrals using an infinite series, one can also readily obtain the GHF-based solutions of the two integrals from the Mathematica software, as will be elaborated later in Chap. 3. However, it is unlikely that one can relate the two infinite series expressed in terms of the hypergeometric series (see Appendix A for detail) with those adopted by Bakhmeteff (1932),Chow (1955, 1957, 1959) in their computation of the VFF or derived by Ramamurthy et al. (2000) using solely the binomial expansion of the rational function. The software **Mathematica** is a computational software program used in many scientific, engineering, mathematical and computing fields, see Wolfram (1996). Except of the **Mathematica** software, the GHF-based GVF solutions discussed in this book can be solved by other similar mathematical programs, such as the **Maple** software (Bernardin et al. 2011) and the **Matlab** software (Houcque 2005).

In addition, there are some researchers have tried to directly solve the dynamic GVF Eq. i.e., (1.6), instead of (1.41), without using the hydraulic exponents M and N, under some approximation techniques. Dubin (1999) ever proposed an approximate semi-analytical solution for GVF in rectangular channels using the Manning formula. Vatankhah (2010) proposed an analytical solution for GVF in triangular channels using the Manning formula. Vatankhah and Easa (2011) derived a semi-analytical solution for GVF in general rectangular channels using the Manning formula. Vatankhah (2011) presented a direction integration method to derive a semi-analytical solution of the Manning-based GVF profiles in parabolic. Vatankhah (2012) also derived a direction integration method to derive a semi-analytical solution of the Manning-based GVF profiles in trapezoidal channels. Note that these kinds of semi-analytical solutions, similar to the above-mentioned ETF-based solutions, are lengthy and disorganized.

Table 2.2 The ETF-based solution of the dimensionless GVF profile for flow in a wide sustaining channel with hydraulic exponents $M = 3$ and $N = 10/3$

Equation: $x_* = u - \int \frac{1}{1-u^{10/3}} du + \lambda^3 \int \frac{u^{1/3}}{1-u^{10/3}} + \text{Const.}.$

Solution (for $0 \leq u < 1$ or $u > 1$):

$$
\begin{aligned}
x_* = u - \frac{3}{40} \Bigg\{ & -2\sqrt{2\left(5+\sqrt{5}\right)} \arctan\left[\frac{1+\sqrt{5}-4u^{1/3}}{\sqrt{10-2\sqrt{5}}}\right] - 2\sqrt{10-2\sqrt{5}} \arctan\left[\frac{1-\sqrt{5}+4u^{1/3}}{\sqrt{10+2\sqrt{5}}}\right] \\
& - 2\sqrt{10-2\sqrt{5}} \arctan\left[\frac{-1+\sqrt{5}+4u^{1/3}}{\sqrt{10+2\sqrt{5}}}\right] + 2\sqrt{10-2\sqrt{5}} \arctan\left[\frac{1+\sqrt{5}+4u^{1/3}}{\sqrt{10-2\sqrt{5}}}\right] - 4\ln\left[-1+u^{1/3}\right] + 4\ln\left[1+u^{1/3}\right] \\
& + (1+\sqrt{5})\ln\left[1-\frac{1}{2}(-1+\sqrt{5})u^{1/3}+u^{2/3}\right] - (1+\sqrt{5})\ln\left[1+\frac{1}{2}(-1+\sqrt{5})u^{1/3}+u^{2/3}\right] \\
& + (-1+\sqrt{5})\ln\left[1-\frac{1}{2}(1+\sqrt{5})u^{1/3}+u^{2/3}\right] - (-1+\sqrt{5})\ln\left[1+\frac{1}{2}(1+\sqrt{5})u^{1/3}+u^{2/3}\right]\Bigg\} \\
+ \frac{3\lambda^3}{40} \Bigg\{ & -2\sqrt{2\left(5+\sqrt{5}\right)} \arctan\left[\frac{1+\sqrt{5}-4u^{1/3}}{\sqrt{10-2\sqrt{5}}}\right] - 2\sqrt{10-2\sqrt{5}} \arctan\left[\frac{1-\sqrt{5}+4u^{1/3}}{\sqrt{10+2\sqrt{5}}}\right] \\
& - 2\sqrt{10-2\sqrt{5}} \arctan\left[\frac{-1+\sqrt{5}+4u^{1/3}}{\sqrt{10+2\sqrt{5}}}\right] - 2\sqrt{10-2\sqrt{5}} \arctan\left[\frac{1+\sqrt{5}+4u^{1/3}}{\sqrt{10-2\sqrt{5}}}\right] - 4\ln\left[-1+u^{1/3}\right] \\
& - 4\ln\left[1+u^{1/3}\right] - (-1+\sqrt{5})\ln\left[1-\frac{1}{2}\left(-1+\sqrt{5}\right)u^{1/3}+u^{2/3}\right] - (-1+\sqrt{5})\ln\left[1+\frac{1}{2}\left(-1+\sqrt{5}\right)u^{1/3}+u^{2/3}\right] \\
& + (1+\sqrt{5})\ln\left[1-\frac{1}{2}(1+\sqrt{5})u^{1/3}+u^{2/3}\right] + (1+\sqrt{5})\ln\left[1+\frac{1}{2}\left(1+\sqrt{5}\right)u^{1/3}+u^{2/3}\right]\Bigg\} + \text{Const.}.
\end{aligned}
$$

2.8 Summary

The direct-integration method is a conventional method used to analytically solve the GVF equation. This chapter has introduced the conventional methods used to find the direct integration solutions of the GVF equation by using the varied-flow functions (VFF) and the elementary transcendental functions (ETF). The GVF equation is usually normalized using the normal depth, h_n before it can be analytically solved by the direct integration method. The hydraulic exponents M and N, and the ratio of the critical depth to normal depth (h_c/h_n) are the three key parameters in the normalized GVF equation. For simplicity, M and N are usually treated as constant, thus averting the impasse in integrating directly the GVF equation. The GVF solution by the Bakhmeteff-Chow procedure with VFF-tables is reviewed. These VFF-tables needed in the direct-integration method has a drawback caused by the imprecise interpolation of the VFF-values. The ETF-based solutions are lengthy and disorganized, except for the special case of $M = N = 3$ (Bresse's solution). To overcome these drawbacks, we will use the Gaussian hypergeometric functions (GHF) to analytically solve the GVF equation without recourse to the VFF tables and the lengthy elementary transcendental functions, as shown in the following chapters.

References

Allen J, Enever KJ (1968) Water surface profiles in gradually varied open-channel flow. Proc Inst Civil Eng 41:783–811

Bakhmeteff BA (1932) Hydraulics of Open Channels. McGraw-Hill, New York

Bernardin L, Chin P, DeMarco P, Geddes KO, Hare DEG, Heal KM, Labahn G, May JP, McCarron J, Monagan MB, Ohashi D, Vorkoetter SM (2011) Maple programming guide. Maplesoft, Waterloo Maple Inc., Canada

Bresse J (1860) Cours de mécanique appliqué, 2e partie edition. hydraulique. Mallet-Bachelier, Paris

Chow VT (1957) Closure of discussions on integrating the equation of gradually varied flow. J Hydraul Div ASCE 83(1):9–22

Chow VT (1959) Open-channel hydraulics. McGraw-Hill, New York

Chow VT (1981) Hydraulic exponents. J Hydraul Div ASCE 107(HY11):1489–1499

Chow VT (1955) Integrating the equation of gradually varied flow. Proceedings Paper No. 838. ASCE 81:1–32

Dubin JR (1999) On gradually varied flow profiles in rectangular open channels. J Hydraul Res 37(1):99–106

Gibson AH (1934) Hydraulics and its applications, 4th edn. Constable and Co., London

Gill MA (1976) Exact solution of gradually varied flow. J Hydraul Div ASCE 102(HY9):1353–1364

Gunder DF (1943) Profile curves for open-channel flow. ASCE Trans 108:481–488

Houcque D (2005) Introduction to Matlab for engineering student. Northwestern University, USA

Jan CD, Chen CL (2012) Use of the Gaussian hypergeometric function to solve the equation of gradually-varied flow. J Hydrol 456–457:139–145

Jan CD, Chen CL (2013) Gradually varied open-channel flow profiles normalized by critical depth and analytically solved by using Gaussian hypergeometric functions. HESS 17:973–987

Keifer CJ, Chu HH (1955) Backwater functions by numerical integration. ASCE Trans 120:429–442

Kirpich PZ (1948) Dimensionless constants for hydraulic elements of open-channel cross-sections. Civil Eng 18(10):47–49

Kumar A (1978) Integral solutions of the gradually varied equation for rectangular and triangular channels. Proc Inst Civil Eng 65(2):509–515

Lee M (1947) Steady gradually varied flow in uniform channels on mild slopes. Ph.D. thesis, Department of Civil Engineering, University of Illinois at Urbana-Champaign

Mononobe N (1938) Back-water and drop-down curves for uniform channels. ASCE Trans 103:950–989

Patil RG, Diwanji VN, Khatsuria RM (2001) Integrating equation of gradually varied flow. J Hydraul Eng ASCE 127(7):624–625

Patil RG, Diwanji VN, Khatsuria RM (2003) Closure to integrating equation of gradually varied flow by R. G. Patil, V. N. Diwanji, R. M. Khatsuria. J Hydraul Eng ASCE 129(1):78–78

Pickard WF (1963) Solving the equations of uniform flow. J Hydraul Div ASCE 89(HY4):23–37

Ramamurthy AS, Saghravani SF, Balachandar R (2000) A direct integration method for computation of gradually varied flow profiles. Can J Civil Eng 27(6):1300–1305

Shivareddy MS, Viswanadh GK (2008) Hydraulic exponents for a D-shaped tunnel section. Int J Appl Eng Res 3(8):1019–1028

Srivastava R (2003) Discussion of "integrating equation of gradually varied flow" by R. G. Patil, V. N. Diwanji, and R. M. Khatsuria. J Hydraul Eng ASCE 129(1):77–78

Subramanya K (2009) Flow in open channels, 3rd edn. McGraw-Hill, Singapore

Subramanya K (2009) Flow in open channels, 3rd edn. McGraw-Hill, Singapore

Subramanya K, Ramamurthy AS (1974) Flow profile computation by direct integration. J Hydraul Div ASCE 100(HY9):1307–1311

Vatankhah AR, Easa SM (2011) Direction integration of Manningbased GVF flow equation. Water Manage ICE 164(5):257–264

Vatankhah AR (2010) Analytical integration of the equation of gradually varied flow for triangular channels. Flow Measur Instrum 21(4):546–549

Vatankhah AR (2011) Direction integration of gradually varied flow equation in parabolic channels. Flow Measur Instrum 22(3):235–234

Vatankhah AR (2012) Direction integration of Manning-based GVF equation in trapezodial channels. J Hydrol Eng ASCE 17(3):455–462

Venutelli M (2004) Direct integration of the equation of gradually varied flow. J Irrig Drainage Eng ASCE 130(1):88–91

Von Seggern ME (1950) Integrating the equation of nonuniform flow. ASCE Trans 115:71–88

Wolfram S (1996) The mathematica book, 3rd edn. Wolfram Media and Cambridge University Press, Champaign

Woodward SM, Posey CJ (1941) Hydraulics of steady flow in open channels. Wiley, New York

Zaghloul NA (1990) A computer model to calculate varied flow functions for circular channels. Adv Eng Softw 12(3):106–122

Zaghloul NA (1992) Gradually varied flow in circular channels with variable roughness. Adv Eng Softw 15(1):33–42

Zaghloul NA (1998) Hydraulic exponents M and N for gravity flow pipes. Adv Water Resour 21(3):185–191

Zaghloul NA, Anwar MN (1991) Numerical integration of gradually varied flow in trapezoidal channel. Comput Method Appl Mech Eng 88:259–272

Chapter 3
Normal-Depth-Based Dimensionless GVF Solutions Using the Gaussian Hypergeometric Function

3.1 Introduction

The literature survey as shown in Chap. 2 recaps two general ways (i.e., VFF-based and ETF-based methods) to obtain the exact solutions of the GVF equation through the two integrals of a proper fraction. All previous researchers assumed constant hydraulic exponents, M and N, in their direct integration of the two integrals. However, dissatisfaction with what they have achieved under such a limited condition imposed on the values of M and N has motivated us to remove the inaccuracy in using the imprecise varied-flow function (VFF) table to evaluate the two integrals. The VFF needed in the typical direct-integration method has a drawback caused by the imprecise interpolation of the VFF-values by using the VFF tables, as mentioned in Chap. 2. This chapter is directed to integrate directly the two integrals without recourse to the VFF table. Among the optimum ways to achieve an accurate integration, it is believed to have the integrated results expressed in terms of the *Gaussian hypergeometric functions* (GHF), i.e., an infinite series resulting from the integration of two integrals of a proper fraction. The GHF-based solution so obtained can be validated using the other analytical solution, which should be obtainable from a partial-fraction expansion of the two integrals into two finite sets of partial fractions, each term of which can be integrated separately using the elementary transcendental functions (ETF). The primary objective of this chapter is thus to show that the GHF upon validation of the GHF-based solution can henceforth replace the VFF table. By implication, one can further aim to prove that the VFF table will be no longer needed in the evaluation of the two integrals.

Because, nowadays, some computational software programs, such as the softwares of **Mathematica**, **Maple** and **Matlab**, can accurately output both GHF-based and ETF-based solutions, we will use it wherever possible in this study so as to facilitate the acquisition of both solutions for various N-values. For validation, we can show that both solutions for a specified N-value (which should be limited to a rational number for the ETF-based solution, but can be any real number for the GHF-based solution) are identical. Each N-value represents the mth power of the power-law

C.-D. Jan, *Gradually-varied Flow Profiles in Open Channels*, Advances in Geophysical and Environmental Mechanics and Mathematics, DOI: 10.1007/978-3-642-35242-3_3, © Springer-Verlag Berlin Heidelberg 2014

flow resistance formula which will be incorporated in the formulation of the GVF equation. Upon normalization of the GVF equation based on the normal depth h_n, we will solve it using the GHF and proceed to prove and/or analyze:

- the properties of the GHF-based solution that can describe the GVF profiles on all types of channel slopes except on the nonsustaining (horizontal or adverse) slopes;
- the slopes of flow profiles which vary with h_c/h_n, M, and N;
- the singularties of the rational function representing the slopes of the C1 and C3 profiles at the dimensionless normal depth (i.e. unity);
- the points of inflection on the M1 and M3 profiles; and
- the curvature of all types of the GVF profiles.

All other related issues with regard to the normalization of the GVF equation based on h_c, such as how to acquire the GHF-based solutions thereof and the subsequent proof and/or analysis of the features and properties of the h_c-based solutions, are addressed and resolved in Chap. 4.

3.2 Normalization of the GVF Equation

The dynamic equation of GVF flow in a channel, without the assumption of small channel slope, has been shown in (1.41). For the convenience of discussion, we rewrite (1.41) herein

$$\frac{dh}{dx} = S_{0*}\frac{1-(h_n/h)^N}{1-(h_c/h)^M}, \tag{3.1}$$

where $S_{0*} = S_0/\cos\theta = \tan\theta$. For the case of small channel slope, $\cos\theta \approx 1$ and $S_{0*} \approx S_0 = \sin\theta$.

It is expected that the solution of (3.1), if obtainable, can be expressed in a form of h as an implicit function of x with S_{0*}, h_n, h_c, M, and N as parameters. However, before solving (3.1), we can rearrange it into a dimensionless form, which will facilitate the solution thereof. While keeping M and N intact, we can combine the primary variables, x and h, with length parameters to create, respectively, their dimensionless counterparts. Thus, via rearrangement, we can reduce the number of the parameters appearing in (3.1) before the dimensionless form of (3.1) is solved. Besides, rearranging (3.1) into a dimensionless form can by itself disclose the unique dimensionless parameter, h_c/h_n (or its reciprocal, h_n/h_c), which characterizes GVF profiles in sustaining channels, such as those on horizontal ($h_c/h_n = 0$, as $h_n \to \infty$ rather than $h_c \to 0$), mild ($0 < h_c/h_n < 1$), critical ($h_c/h_n = 1$), and steep ($h_c/h_n > 1$) slopes. GVF profiles in adverse channels can also be characterized by h_c/h_n, but h_n used in defining h_c/h_n for GVF profiles in adverse channels is fictitious, as analyzed in the next chapter. We will later explain why GVF profiles based on h_c/h_n should be classified slightly differently from those of Chow (1959).

There are two ways in which x and h can be normalized: One is based on h_n, as adopted by many researchers, such as Chow (1959) and others, and the other on

h_c, as treated exclusively by Chen and Wang (1969), who replaced h_n by h_c on the ground that h_n becomes infinite and/or undefined for flow in horizontal and adverse channels. In this chapter, however, we are confined to normalize x and h based on h_n only and proceed to solve the dimensionless form of (3.1) based on h_n. It is more involved to normalize x and h based on h_c than based on h_n; therefore, we cover the normalization of (3.1) based on h_c in Chap. 4, and henceforth in this paper focus on the normalization of (3.1) based on h_n.

To normalize x and h in (3.1) based on h_n, let us introduce two dimensionless variables u and x_*, and one dimensionless parameter λ as follows

$$u = \frac{h}{h_n}, \tag{3.2}$$

$$x_* = \frac{x S_{0*}}{h_n}, \tag{3.3}$$

$$\lambda = \frac{h_c}{h_n}. \tag{3.4}$$

Then we rearrange the expression of (3.1) in terms of u and x_* as

$$\frac{du}{dx_*} = \frac{u^N - 1}{u^N - \lambda^M u^{N-M}}. \tag{3.5}$$

Equation (3.5) is a dimensionless form of (3.1), describing a dimensionless GVF profile with $\lambda(= h_c/h_n)$, M, and N as parameters. The characteristic length ratio λ is the primary parameter, which classifies GVF under study into those in mild, critical, and steep channels according as λ is less than, equal to, and larger than unity, respectively. By and large, M and N are functions not only of h (or u in case of the dimensionless form thereof), but also of channel geometry and roughness. However, with the present knowledge in mathematics, it is unlikely that one can solve (3.5) analytically unless M and N are treated as constants. Insomuch that, we henceforth assume the invariance of both M and N with u and confine the solutions of (3.5) to a case in which the values of M and N do not vary with u within a given channel geometry.

3.3 GVF Solutions by Using Gaussian Hypergeometric Functions

To integrate (3.5) directly and express its solution in terms of the Gaussian hypergeometric functions (GHF), we have three options to arrange the two integrals of a proper fraction, which will be formed from (3.5). In principle, the GHF-based solutions obtained by adding the two integrals together using one of the three options are presumed to be identical. It is this presumption which paves the way to establish a general functional identity relation (or often called a recurrence formula) between the two contiguous GHF, as can readily be proved by use of induction in (B.3) of

Appendix B. The recurrence formula so established between the two contiguous GHF is valuable in linking two unlikely related GHF. Roles of such a recurrence formula playing in the conversion of the two GHF-based solutions, among other applications, will be elaborated later.

3.3.1 An Alternative Form of (3.5) for $|u| > 1$

Since the solutions of (3.5), if expressed in terms of the GHF, are subject to the convergence criterion of the GHF, i.e., $|u| < 1$, an alternative form of (3.5) has yet to be formulated and its solutions expressed in terms of the reconciled GHF, which is valid for $|u| > 1$. In the following, we firstly derive such an alternative form of (3.5) and proceed to arrange two integrals, which can be formed from both (3.5) and its alternative form in two ways. Secondly, upon derivation of the respective GHF-based solutions from (3.5) and its alternative form (i.e., (3.7)), we can establish two recurrence formulas between a set of the two corresponding contiguous GHF. Thirdly, we verify the identity of such two recurrence formulas by induction, as conducted in Appendix B. The two recurrence formulas will later be used to prove the identity of the four parametrically different, yet mutually related GHF-based solutions of (3.5) and its alternative form.

To derive the alternative form of (3.5) in order for its GHF-based solutions to be valid for $|u| > 1$, we introduce a new variable w, defined as

$$w = \frac{1}{u}. \tag{3.6}$$

Equation (3.5) on substitution of $u = w^{-1}$ and $du = -w^{-2}dw$ yields

$$\frac{dw}{dx_*} = \frac{1 - w^N}{-w^{-2} + \lambda^M w^{M-2}}. \tag{3.7}$$

Obviously, the analytical solutions of (3.7), if expressed in terms of the GHF, are convergent for $|w| < 1$, which on substitution of $w = u^{-1}$ yields $|u| > 1$, i.e., the reciprocal of the convergence criterion of the GHF, $|u| < 1$, imposed on the GHF-based solutions of (3.5).

3.3.2 Feasible Arrangement of Two Integrals

(1) For $|u| < 1$:

Rearranging (3.5) into its reciprocal form yields

$$\frac{dx_*}{du} = \frac{-u^N + \lambda^M u^{N-M}}{1 - u^N}. \tag{3.8}$$

The right-hand side of (3.8) is a rational function of u, which by the process of long division can be expressed in the form of a polynomial plus a proper fraction in two ways before we integrate them as follows: One way is to divide the first term of the numerator on the right-hand side of (3.8) by the denominator, i.e.,

$$\frac{dx_*}{du} = \frac{-u^N + \lambda^M u^{N-M}}{1-u^N} = 1 - \frac{1}{1-u^N} + \frac{\lambda^M u^{N-M}}{1-u^N}. \tag{3.9}$$

The other way is to divide the second term of the numerator on the right-hand side of (3.8) by the denominator, i.e.,

$$\frac{dx_*}{du} = \frac{-u^N + \lambda^M u^{N-M}}{1-u^N} = -\lambda^M u^{-M} - \frac{u^N}{1-u^N} + \frac{\lambda^M u^{-M}}{1-u^N}. \tag{3.10}$$

The integrations of (3.8)–(3.10) with respect to du, respectively, yield

$$x_* = -\int_0^u \frac{u^N}{1-u^N} du + \lambda^M \int_0^u \frac{u^{N-M}}{1-u^N} du + \text{Const.}, \tag{3.11}$$

$$x_* = u - \int_0^u \frac{1}{1-u^N} du + \lambda^M \int_0^u \frac{u^{N-M}}{1-u^N} du + \text{Const.}, \tag{3.12}$$

$$x_* = -\lambda^M \frac{u^{1-M}}{1-M} - \int_0^u \frac{u^N}{1-u^N} du + \lambda^M \int_0^u \frac{u^{-M}}{1-u^N} du + \text{Const.}, \tag{3.13}$$

in which "Const." is the constant of integration. As mentioned earlier, three GHF-based solutions to be obtained from (3.11)–(3.13) should be identical. Therefore, equating any two of such three GHF-based solutions will pave the way to establish two functional identical relations of two contiguous GHF (as will be later shown in (3.32) and (3.33)). It will be found later that these two functional identity relations are only the special cases of a recurrence formula between two contiguous GHF, as shown in (3.34) later.

(2) For $| u |> 1$:

Likewise, we can rearrange (3.7) into its reciprocal form as

$$\frac{dx_*}{dw} = \frac{-w^{-2} + \lambda^M w^{M-2}}{1-w^N}. \tag{3.14}$$

By the same token, we can similarly express the right-hand side of (3.14) in the form of a polynomial plus a proper fraction in two ways as follows: One way is to divide the first term of the numerator on the right-hand side of (3.14) by the denominator, i.e.,

$$\frac{dx_*}{dw} = \frac{-w^{-2} + \lambda^M w^{M-2}}{1-w^N} = -w^{-2} - \frac{w^{N-2}}{1-w^N} + \frac{\lambda^M w^{M-2}}{1-w^N}. \tag{3.15}$$

The other way is to divide the second term of the numerator on the right-hand side of (3.14) by the denominator, i.e.,

$$\frac{dx_*}{dw} = \frac{-w^{-2} + \lambda^M w^{M-2}}{1 - w^N} = \lambda^M w^{M-2} - \frac{w^{-2}}{1 - w^N} + \frac{\lambda^M w^{N+M-2}}{1 - w^N}. \tag{3.16}$$

Integrations of (3.14)–(3.16) with respect to dw, respectively, yield the following to equations,

$$x_* = -\int_0^w \frac{w^{-2}}{1 - w^N} dw + \lambda^M \int_0^w \frac{w^{N-2}}{1 - w^N} dw + \text{Const.}, \tag{3.17}$$

$$x_* = w^{-1} - \int_0^w \frac{w^{N-2}}{1 - w^N} dw + \lambda^M \int_0^w \frac{w^{M-2}}{1 - w^N} dw + \text{Const.}, \tag{3.18}$$

$$x_* = \frac{\lambda^M w^{M-1}}{M - 1} - \int_0^u \frac{w^{-2}}{1 - w^N} dw + \lambda^M \int_0^u \frac{w^{N+M-2}}{1 - w^N} dw + \text{Const.}. \tag{3.19}$$

Likewise, two GHF-based solutions to be obtained from (3.17)–(3.19) are identical. Analogous to the results obtained from (3.11) to (3.13), equating any two of such three GHF-based solutions will establish another two recurrence formulas between two contiguous GHF.

3.3.3 Gaussian Hypergeometric Functions

Prior to find the analytical solutions of the dimensionless equations for the gradually-varied flow in open channels by using the Gaussian hypergeometric function (GHF), we will briefly introduce the definition of GHF and some relations here first. More detailed information about GHF is indicated in Appendix A. The GHF can be expressed as an infinite series and symbolized in the form of ${}_2F_1(a, b; c; z)$ as shown in the books of Abramowitz and Stegun (1972), Korn and Korn (1961); Luke (1975), Pearson (1974), and Seaborn (1991), among others. Olde Daalhuis (2010) suggested that the form ${}_2F_1(a, b; c; z)$ can be safely changed to the more friendly form $F(a, b; c; z)$.

The definition of GHF can be written as

$${}_2F_1(a, b; c; z) = \frac{\Gamma(c)}{\Gamma(a)\Gamma(b)} \sum_{k=0}^{\infty} \frac{\Gamma(a + k)\Gamma(b + k)}{\Gamma(c + k)k!} z^k, \tag{3.20}$$

in which $\Gamma(a)$, $\Gamma(b)$ and $\Gamma(c)$ are Gamma functions; a, b, and c are the function parameters and z is the variable. The infinite series in Eq. (3.20) is convergent for arbitrary a, b, and c, provided that c is neither a negative integer nor zero and that a or b is not negative integers for real $-1 < z < 1$ (or $|z| < 1$), and for $z = \pm 1$

if $c > a + b$ (Olde Daalhuis 2010). Equation (3.20) has the symmetry property ${}_2F_1(a, b; c; z) = {}_2F_1(b, a; c; z)$.

As presented in the next section, the general solutions of the infinite integrals related with the GVF profiles in this book, if expressed in terms of GHF, are all in following form

$$\int \frac{u^{\phi}}{1-u^N}du = \frac{u^{\phi+1}}{\phi+1}\,{}_2F_1\left(1, \frac{\phi+1}{N}; \frac{\phi+1}{N}+1; u^N\right) + \text{Const.} \tag{3.21}$$

The functional parameters a, b, and c of the GHF in the right hand side of (3.21) are $a = 1$, $b = (\phi + 1)/N$, and $c = [(\phi + 1)/N + 1]$, respectively, with a functional variable $z = u^N$. For this special case of the GHF-based solutions, the functional parameters of GHF have two specified relations, $a = 1$ and $c = b + 1$. For this special case, Eq. (3.20) can be written as

$${}_2F_1(1, b; b+1; z) = b\sum_{k=0}^{\infty}\frac{z^k}{b+k}. \tag{3.22}$$

The GHF used in the solutions of GVF profiles herein are all in the form of (3.22). Since the first argument of the GHF in (3.22) is always unity, the second and third arguments differ in one unity, and the fourth argument is a variable only, we can express the GHF in a simpler form. We shall write $g(b, z)$ instead of ${}_2F_1(1, b; b + 1; z)$, to reduce the related equations to shorter expressions, thus facilitating the reading of the manuscript, as Jan and Chen (2012, 2013) did. Therefore, we will use $g(b, z)$ instead of ${}_2F_1(1, b; b+1; z)$ herein to represent GHF when we write out the GHF-based solutions for GVF profiles. The expansion of $g(b, z)$ can be expressed as

$$g(b, z) = b\sum_{k=0}^{\infty}\frac{z^k}{b+k} = 1 + \frac{bz}{b+1} + \frac{bz^2}{b+2} + \ldots\ldots + \frac{bz^k}{b+k} + \ldots\ldots \tag{3.23}$$

3.3.4 The GHF-Based Solutions of GVF Equation

The complete GHF-based solutions of the dimensionless GVF equation are obtained in the following by executing the integration of the two integrals in (3.11)–(3.13) for $| u |< 1$ as well as those in (3.17)–(3.19) for $| w |< 1$ (or $| u |> 1$ by substitution of $w = u^{-1}$ into its solution). It is noted that either a GHF-based solution excludes the point where $u = 1$ or $w = 1$ because the GHF diverges at $u = 1$ or $w = 1$. Physically, this point corresponds to a theoretical limit where the GVF profile runs parallel with the channel bottom, i.e., at $x_* = \pm\infty$.

(1) For $| u |< 1$:

For $| u |< 1$, we can find from the Mathematica software (or the Maple software) the following GHF-based solutions for the four indefinite integrals in (3.11)–(3.13), namely,

$$\int \frac{1}{1-u^N}\,du = ug\left(\frac{1}{N}, u^N\right) + \text{Const.}, \tag{3.24}$$

$$\int \frac{u^{N-M}}{1-u^N}\,du = \frac{u^{N-M+1}}{N-M+1}\,g\left(\frac{N-M+1}{N}, u^N\right) + \text{Const.}, \tag{3.25}$$

$$\int \frac{u^{N}}{1-u^N}\,du = \frac{u^{N+1}}{N+1}g\left(\frac{N+1}{N}, u^N\right) + \text{Const.}, \tag{3.26}$$

$$\int \frac{u^{-M}}{1-u^N}\,du = \frac{u^{-M+1}}{-M+1}\,g\left(\frac{-M+1}{N}, u^N\right) + \text{Const.}. \tag{3.27}$$

According to the above relations, we can summarize them in a general functional form for the solution of the integral in terms of the GHF as

$$\int \frac{u^{\phi}}{1-u^N}\,du = \frac{u^{\phi+1}}{\phi+1}g\left(\frac{\phi+1}{N}, u^N\right) + \text{Const.}, \tag{3.28}$$

in which the parameter (namely exponent) ϕ is a real number; $\phi = 0, N-M, N, -M$ for (3.24)–(3.27), respectively. The symbol $g(b, z)$ expressed in the above equations is the simplified expression of the Gaussian hypergeometric function ${}_2F_1(1, b; b+1; z)$, with $b = (\phi+1)/N$ and $z = u^N$ herein. The existence of the above relation for the indefinite integral can be proved by using the commercial Mathematica software (Wolfram 1996).

Replacing the two integral terms in each of (3.11)–(3.13) using the GHF-based integration results as shown in (3.24)–(3.27), we can obtained three possible expressions of the GHF-based dimensionless solutions for GVF profiles under the condition $|\, u \,| < 1$, i.e.,

$$x_* = -\frac{u^{N+1}}{N+1}g\left(\frac{N+1}{N}, u^N\right) + \frac{\lambda^M u^{N-M+1}}{N-M+1}g\left(\frac{N-M+1}{N}, u^N\right) + \text{Const.}, \tag{3.29}$$

$$x_* = u\left[1-g\left(\frac{1}{N}, u^N\right)\right] + \frac{\lambda^M u^{N-M+1}}{N-M+1}g\left(\frac{N-M+1}{N}, u^N\right) + \text{Const.}, \tag{3.30}$$

$$x_* = -\frac{u^{N+1}}{N+1}g\left(\frac{N+1}{N}, u^N\right) + \frac{\lambda^M u^{-M+1}}{-M+1}\left[-1+g\left(\frac{-M+1}{N}, u^N\right)\right] + \text{Const.}. \tag{3.31}$$

The above three equations for the GVF solutions are valid only for $|\, u \,| < 1$. Because these three GHF-based solutions so obtained from (3.11)–(3.13), should be identical, equating any two of the three equations (i.e., (3.29)–(3.31)) establishes the following two identities.

$$g\left(\frac{1}{N}, u^N\right) = 1 + \frac{u^N}{N+1} g\left(\frac{N+1}{N}, u^N\right), \tag{3.32}$$

$$g\left(\frac{-M+1}{N}, u^N\right) = 1 + \frac{(-M+1)u^N}{N-M+1} g\left(\frac{N-M+1}{N}, u^N\right). \tag{3.33}$$

It is noted that (3.32) and (3.33) are two special cases of the general functional identity relation (so-called recurrence formulas) of the two contiguous GHF, namely

$$g(b, z) = 1 + \frac{bz}{b+1} g(b+1, z), \tag{3.34}$$

in which the variable $z = u^N$ herein, and the parameters $b = 1/N$ and $(-M + 1)/N$ for the cases of (3.32) and (3.33), respectively. This general functional identity relation is proved as shown in Appendix B.

(2) For $| u |> 1$:

For $| u |> 1$ (or $| w |< 1$ by substitution of $u = w^{-1}$ into $| u |> 1$), we can also find from the Mathematica software the GHF-based solutions of the following four indefinite integrals in (3.17)–(3.19), namely,

$$\int \frac{w^{N-2}}{1-w^N} dw = \frac{w^{N-1}}{N-1} g\left(\frac{N-1}{N}, w^N\right) + \text{Const.}, \tag{3.35}$$

$$\int \frac{w^{M-2}}{1-w^N} dw = \frac{w^{M-1}}{M-1} g\left(\frac{M-1}{N}, w^N\right) + \text{Const.}, \tag{3.36}$$

$$\int \frac{w^{-2}}{1-w^N} dw = -w^{-1} g\left(-\frac{1}{N}, w^N\right) + \text{Const.}, \tag{3.37}$$

$$\int \frac{w^{M+N-2}}{1-w^N} dw = \frac{w^{N+M-1}}{N+M-1} g(\frac{N+M-1}{N}, w^N) + \text{Const.}. \tag{3.38}$$

According to the above relations, we can summarize them in a general functional form for the solution of the indefinite integral in terms of GHF as shown in (3.28), in which the variable u^N is replaced by w^N, and the parameter $\phi = N-2, M-2, -2$ and $M+N-2$ for (3.35)–(3.38), respectively.

Replacing the two integral terms in (3.17) by using (3.36) and (3.37) yields the GVF solution in terms of GHF as

$$x_* = w^{-1} g\left(\frac{-1}{N}, w^N\right) + \frac{\lambda^M w^{M-1}}{M-1} g\left(\frac{M-1}{N}, w^N\right) + \text{Const.}. \tag{3.39}$$

Equation (3.18), on substitution of (3.35) and (3.36), yields the GVF solution in terms of GHF, i.e.,

$$x_* = w^{-1}\left[1 - \frac{w^N}{N-1} g\left(\frac{N-1}{N}, w^N\right)\right] + \frac{\lambda^M w^{M-1}}{M-1} g\left(\frac{M-1}{N}, w^N\right) + \text{Const.}. \tag{3.40}$$

Likewise, (3.19) on substitution of (3.37) and (3.38) yields

$$x_* = w^{-1} g\left(-\frac{1}{N}, w^N\right) + \frac{\lambda^M w^{M-1}}{M-1} \times \left[1 + \frac{(M-1)w^N}{N+M-1} g\left(\frac{N+M-1}{N}, w^N\right)\right] + \text{Const.}. \tag{3.41}$$

The above three equations (i.e., (3.39)–(3.41)) are valid for $| \ w \ | < 1$. To express these three equations in terms of u, by substituting $w = u^{-1}$ into the equation yields the following three equations

$$x_* = u g\left(\frac{-1}{N}, u^{-N}\right) + \frac{\lambda^M u^{-M+1}}{M-1} g\left(\frac{M-1}{N}, u^{-N}\right) + \text{Const.}, \tag{3.42}$$

$$x_* = u\left[1 - \frac{u^{-N}}{N-1} g\left(\frac{N-1}{N}, u^{-N}\right)\right] + \frac{\lambda^M u^{-M+1}}{M-1} g\left(\frac{M-1}{N}, u^{-N}\right) + \text{Const.}, \tag{3.43}$$

$$x_* = u g\left(-\frac{1}{N}, u^{-N}\right) + \frac{\lambda^M u^{-M+1}}{M-1} \times \left[1 + \frac{(M-1)u^{-N}}{N+M-1} g\left(\frac{N+M-1}{N}, u^{-N}\right)\right] + \text{Const.}. \tag{3.44}$$

The above three GHF-based GVF solutions, (3.42) through (3.44) are valid only for $| \ u \ | > 1$. By the same token, Eqs. (3.39)–(3.41) should be identical. Consequently, equating these three equations reveals the existence of the following two recurrence formulas.

$$g\left(-\frac{1}{N}, w^N\right) = 1 - \frac{w^N}{N-1} g\left(\frac{N-1}{N}, w^N\right), \tag{3.45}$$

$$g\left(\frac{M-1}{N}, w^N\right) = 1 + \frac{(M-1)w^N}{N+M-1} g\left(\frac{N+M-1}{N}, w^N\right). \tag{3.46}$$

Alternatively, Eqs. (3.42)–(3.44) should be identical. Comparison of these three equations also yields the existence of two analogous recurrence formulas as follows:

$$g\left(-\frac{1}{N}, u^{-N}\right) = 1 - \frac{u^{-N}}{N-1} g\left(\frac{N-1}{N}, u^{-N}\right), \tag{3.47}$$

$$g\left(\frac{M-1}{N}, u^{-N}\right) = 1 + \frac{(M-1)u^{-N}}{N+M-1} g\left(\frac{N+M-1}{N}, u^{-N}\right). \tag{3.48}$$

Obviously, (3.45) is identical to (3.47), and (3.46) is identical to (3.48), on substitution of $w = u^{-1}$. Likewise, (3.47) and (3.48) are two special cases of the recurrence formulas of the two contiguous GHF as shown in (3.34), in which the variable $z = u^{-N}$ and the parameter $b = -1/N$ and $(M-1)/N$ for the cases of (3.47) and (3.48), respectively.

Although the absolute value of the variable in the GHF, i.e., $| u |$, has been imposed to derive the solutions of (3.5) for GVF in wide sustaining channels, the complete GHF-based solutions of (3.5) can only cover the physically possible domain of u, i.e. $0 \leq u < \infty$, thus consisting of (3.30) (or (3.29) and (3.31)) in the domain of $0 \leq u < 1$, $x_* = \pm\infty$ at $u = 1$, and (3.43) (or (3.42) and (3.44)) in the domain of $u > 1$. Therefore we can lift the absolute-value restriction imposed on u in expressing the GHF-based solutions of (3.5). Thus, except for $u = 1$, the complete GHF-based solutions of (3.5) should consist of any combination of (3.30) (or (3.29) and (3.31)) in the domain of $0 \leq u < 1$ with (3.43) (or (3.42) and (3.44)) in the domain of $u > 1$. By implication, any of the nine combinations, i.e., (3.30) combined with (3.42), (3.43) or (3.44); (3.31) combined with (3.42), (3.43) or (3.44); and (3.32) combined with (3.42), (3.43) or (3.44), can represent the complete GHF-based solutions of (3.5) for $0 \leq u < \infty$, excluding at $u = 1$. Any differences in the GHF expressions among them can be reconciled by one another through conversion using the recurrence formula, i.e., (3.34). Henceforth in this chapter, we can use solely the combination of (3.30) and (3.43) as typical of the complete solutions of (3.5) spanning the two respective domains of u, i.e., $0 \leq u < 1$ and $u > 1$, as indicated in Table 3.1, unless a circumstance in favor of the other combinations arises otherwise.

3.4 Alternative Method to Get the GHF-Based Solutions

It is worthwhile to point out that we can alternatively get the solution for $| u | > 1$ directly from the solution for $| u | < 1$ through an established relation connecting one GHF in the domain of $| u | < 1$ to two GHF in the domain of $| u | > 1$, such as (C.3) derived in Appendix C. In other words, such a relation enables one to transform the GHF-based solution of (3.5) in the domain of $| u | < 1$, i.e., (3.30), to the expected GHF-based solution of (3.5) in the domain of $| u | > 1$, i.e., (3.43). Though we have so far obtained the GHF-based solution in the domain of $| u | > 1$ from (3.7) (i.e., a transformed form of (3.5) with its variable, w, being expressed as u^{-1} rather than from (3.5), we can prove that (3.43) is also obtainable directly from (3.30) using the

Table 3.1 Equations for eight h_n-based dimensionless GVF profiles in terms of the GHF

Equations for dimensionless GVF profiles expressed by using the GHF with the hydraulic exponents, M and N	Eq. No.
For M1 $(1 < u < \infty)^a$, S1 $(\lambda \leq u < \infty)^b$ and S2 $(1 < u \leq \lambda)$: $x_* = u\left[1 - \frac{u^{-N}}{N-1} g\left(\frac{N-1}{N}, u^{-N}\right)\right] + \frac{\lambda^M u^{-M+1}}{M-1} g\left(\frac{M-1}{N}, u^{-N}\right) + \text{Const.}.$	(3.43)
For M2 $(\lambda \leq u < 1)$, M3 $(0 \leq u < \lambda)$ and S3 $(0 \leq u < 1)$: $x_* = u\left[1 - g\left(\frac{1}{N}, u^N\right)\right] + \frac{\lambda^M u^{N-M+1}}{N-M+1} g\left(\frac{N-M+1}{N}, u^N\right) + \text{Const.}.$	(3.30)
For C1 $(1 < u < \infty)$: $x_* = u\left[1 - \frac{u^{-N}}{N-1} g\left(\frac{N-1}{N}, u^{-N}\right)\right] + \frac{u^{-M+1}}{M-1} g\left(\frac{M-1}{N}, u^{-N}\right) + \text{Const.}.$	(3.50)
For C3 $(0 \leq u < 1)$: $x_* = u\left[1 - g\left(\frac{1}{N}, u^N\right)\right] + \frac{u^{N-M+1}}{N-M+1} g\left(\frac{N-M+1}{N}, u^N\right) + \text{Const.}.$	(3.49)

[a] The dimensionless flow depth $u = h/h_n$. The h_c-based GVF profiles expressed in terms of GHF can describe two more types of GVF profiles on each of the horizontal and adverse slopes than its counterpart based on h_n; see the next chapter

[b] The parameter λ $(= h_c/h_n)$ is less than, equal to and larger than unity for M, C and S curves, respectively

relation, such as (C.3), to connect one GHF in the domain of $|u| < 1$ to two GHF in the domain of $|u| > 1$. It is shown in Appendix C that (C.7) so transformed from (3.30) through (C.3) is identical to (3.43). Likewise, we can also prove that (3.30) is obtainable from (3.43) in the reverse transformation through (C.4) as indicated in (C.10), which is shown to be identical to (3.30). For further details, see Appendix C.

Once we decide to use (3.30) and (3.43) as typical of the complete solutions of (3.5) for $0 \leq u < \infty$, excluding at $u = 1$, what remains to be determined in (3.30) and (3.43) before applying them to practical problems are the functional parameters, M and N, in the GHF-based solutions, and the integration constants in (3.30) and (3.43). In the following, we will determine such functional parameters and integration constants before plotting and analyzing (3.30) and (3.43).

3.5 Classification of GHF-Based Solutions

3.5.1 M, C and S Profiles

We can make classification of GHF-based solutions obtained in the previous section by using the parameter λ ($= h_c/h_n$). The value of λ is less than, equal to and larger than unity for M, C and S profiles, respectively. In contrast to twelve types of the h_c-based GVF profiles, as shown in the next chapter, there exist only eight types of the h_n-based GVF profiles, three on the mild (M) ($h_c/h_n < 1$) slope, two on the critical (C) ($h_c/h_n = 1$) slope, and three on the steep (S) ($h_c/h_n > 1$) slope, for h_n is infinite on the horizontal (H) slope and undefined on the adverse (A) slope. Eight h_n-based GVF profiles on the M, C, and S slopes are, respectively, referred to as M1, M2, and M3; C1 and C3; S1, S2, and S3, and its entire domain of u, i.e., $0 \le u < \infty$, is demarcated according to the classification of Chow (1959) except C2, which is excluded from this chapter for a reason to be justified later. In fact, the eight h_n-based GVF profiles so classified can be described using either (3.30) or (3.43), as tabulated in Table 3.1, except those on the C slope.

For GVF profiles in C slope, they are more specifically described by (3.30) and (3.43) on substitution of $\lambda = 1$ as

$$x_* = u\left[1 - g\left(\frac{1}{N}, u^N\right)\right] + \frac{u^{N-M+1}}{N-M+1} g\left(\frac{N-M+1}{N}, u^N\right) + \text{Const.}, \tag{3.49}$$

$$x_* = u\left[1 - \frac{u^{-N}}{N-1} g\left(\frac{N-1}{N}, u^{-N}\right)\right] + \frac{u^{-M+1}}{M-1} g\left(\frac{M-1}{N}, u^{-N}\right) + \text{Const.}. \tag{3.50}$$

Therefore, (3.49) for $0 < u < 1$ and (3.50) for $u > 1$ are equations describing the C3 and C1 profiles, respectively. In a special case, (3.49) upon substitution of $M = N = 3$ readily reduces to

$$x_* = u + \text{Const.} \tag{3.51}$$

Unlike the M and S profiles that diverge at $u = 1$, the C1 and C3 profiles, as represented, respectively, by (3.50) and (3.49), neither diverge nor converge at $u = 1$ because both equations have a singularity at $u = 1$, except for $M = N = 3$. As shown above, the C1 and C3 profiles in the simplest case of $M = N = 3$ become a continuous straight line without a singularity being created at $u = 1$. This result can readily be deduced from (3.5) on substitution of $h_c/h_n = 1$ and $M = N = 3$. We shall prove the existence of the singularities on the C3 and C1 profiles at $u = 1$ later when the properties of the h_n-based GVF profiles are analyzed.

Table 3.1 shows the summary of the equations for the eight GVF profiles in sustaining channels: Eq. (3.30) for M1 ($1 < u < \infty$), S1 ($\lambda \le u < \infty$), and S2 ($1 < u \le \lambda$); (3.43) for M2 ($\lambda \le u < 1$), M3 ($0 \le u \le \lambda$) and S3 ($0 \le u < 1$); (3.50) for C1 ($1 < u < \infty$); and (3.49) for C3 ($0 \le u < 1$).

3.5.2 An Example of GHF-Based GVF Profiles with Specified Hydraulic Exponents

As mentioned in Sect. 3.4, (3.30) and (3.43) are the typical solutions for the h_n-based dimensionless GVF equation, as shown in (3.5), for $0 \leq u < 1$ and $u > 1$, respectively. For the special case with the hydraulic exponents, $M = 3$ and $N = 10/3$, the equations for dimensionless GVF profiles expressed in terms of GHF can be written as follows:

For M1 ($1 < u < \infty$), S1 ($\lambda \leq u < \infty$), and S2 ($1 < u < \lambda$):

$$x_* = u\left[1 - \frac{3u^{-10/3}}{7} g\left(\frac{7}{10}, u^{-10/3}\right)\right] + \frac{\lambda^3 u^{-2}}{2} g\left(\frac{3}{5}, u^{-10/3}\right) + \text{Const.}. \quad (3.52)$$

For M2 ($\lambda \leq u < 1$), M3 ($0 \leq u \leq \lambda$) and S3 ($0 \leq u < 1$):

$$x_* = u\left[1 - g\left(\frac{3}{10}, u^{10/3}\right)\right] + \frac{3\lambda^3 u^{4/3}}{4} g\left(\frac{2}{5}, u^{10/3}\right) + \text{Const.}. \quad (3.53)$$

For C1 ($1 < u < \infty$) with $\lambda = 1$:

$$x_* = u\left[1 - \frac{3u^{-10/3}}{7} g\left(\frac{7}{10}, u^{-10/3}\right)\right] + \frac{u^{-2}}{2} g\left(\frac{3}{5}, u^{-10/3}\right) + \text{Const.}. \quad (3.54)$$

For C3 ($0 \leq u < 1$) with $\lambda = 1$:

$$x_* = u\left[1 - g\left(\frac{3}{10}, u^{10/3}\right)\right] + \frac{3u^{4/3}}{4} g\left(\frac{2}{5}, u^{10/3}\right) + \text{Const.}. \quad (3.55)$$

The recurrence formulas (3.47) and (3.32) on substitution of $M = 3$ and $N = 10/3$ yield two functional identity relations of two contiguous GHF, respectively, namely,

$$g\left(\frac{-3}{10}, u^{-10/3}\right) = 1 - \frac{3u^{-10/3}}{7} g\left(\frac{7}{10}, u^{-10/3}\right), \quad (3.56)$$

$$g\left(\frac{3}{10}, u^{10/3}\right) = 1 + \frac{3u^{10/3}}{13} g\left(\frac{13}{10}, u^{10/3}\right). \quad (3.57)$$

Substitution of the above two functional identity relations into (3.52) through (3.55), we can get alternative equations for GHF-based GVF profiles with $M = 3$ and $N = 10/3$, as listed in (3.58) through (3.61).

For M1 ($1 < u < \infty$), S1 ($\lambda < u < \infty$), and S2 ($1 < u < \lambda$):

$$x_* = ug\left(\frac{-3}{10}, u^{-10/3}\right) + \frac{\lambda^3 u^{-2}}{2} g\left(\frac{3}{5}, u^{-10/3}\right) + \text{Const.}. \quad (3.58)$$

For M2 ($\lambda \le u < 1$), M3 ($0 \le u \le \lambda$) and S3 ($0 \le u < 1$):

$$x_* = -\frac{3u}{13} g\left(\frac{13}{10}, u^{10/3}\right) + \frac{3\lambda^3 u^{4/3}}{4} g\left(\frac{2}{5}, u^{10/3}\right) + \text{Const.}. \quad (3.59)$$

For C1 ($1 < u < \infty$) with $\lambda = 1$:

$$x_* = ug\left(\frac{-3}{10}, u^{-10/3}\right) + \frac{u^{-2}}{2} g\left(\frac{3}{5}, u^{-10/3}\right) + \text{Const.}. \quad (3.60)$$

For C3 ($0 \le u < 1$) with $\lambda = 1$:

$$x_* = -\frac{3u}{13} g\left(\frac{13}{10}, u^{10/3}\right) + \frac{3u^{4/3}}{4} g\left(\frac{2}{5}, u^{10/3}\right) + \text{Const.}. \quad (3.61)$$

3.5.3 *N*-Values for Fully Rough Flows in Wide Channels

Mathematically, the parameter, N, in the GHF-based solutions, as tabulated in Table 3.1, can be valid for any real number, but in this book we are only interested in the N-values to be determined from (1.40), i.e., $N = 2m + 3$ for fully rough flows, as shown in Table 3.2. Thus, for application in the computation of GVF profiles in wide channels, our choice will be confined to such five N-values listed in Table 3.2 (i.e., $N = 3, 10/3, 17/5, 7/2, 11/3$) for fully rough flows unless specified otherwise for other applications.

3.5.4 GHF-Based Solutions Under Specified Boundary Conditions

Once the N-value is selected, the constant of integration, "Const.", in (3.30) and (3.49) for $0 \le u < 1$ as well as in (3.43) and (3.50) for $u > 1$ can be determined from the prescribed boundary conditions of a problem under study. Because a particular solution of (3.5) is composed of two equations in (3.30) and (3.43) (or (3.49) and (3.50) if $h_c/h_n = 1$), we only need two boundary conditions, one for each equation, to determine "Const." in each equation. For illustration, we plot (3.43 (or (3.50)) and (3.30) (or (3.49)) for $M = 3$ and $N = 10/3$ (a typical value corresponding to the Manning formula) using two boundary conditions set at $(x_*, u) = (-3, 2)$ and $(-4, 0)$, respectively, as shown in Fig. 3.1. We first plot (3.43) and (3.30) for both M

Table 3.2 Relation of N-value and mth power from ($N = 2m + 3$) for fully rough flows

mth power	0	1/6[a]	1/5	1/4[a]	1/3
N-value	3	10/3	17/5	7/2	11/3

[a] The one-sixth power and one-fourth power formulas are often referred to as the Manning and Lacey formulas, respectively (Chen (1991))

and S profiles with h_c/h_n as a parameter, and then draw (3.50) and (3.49) for the C1 and C3 profiles, respectively, with $h_c/h_n = 1$. It should be noted that both (3.43) and (3.30) diverge at $u = 1$, whereas both (3.50) for $u > 1$ and (3.49) for $0 \leq u < 1$ can be proved to have a singularity at $u = 1$, thus becoming undefined, as marked by two small open circles in Fig. 3.1.

Strictly speaking, we only need two boundary conditions, one for each of (3.43) and (3.30) (or (3.50) and (3.49) if $h_c/h_n = 1$), to determine "Const." in each equation, as stated above. However, we may purposely use three rather than two boundary conditions corresponding to the three channel zones, as classified in Table 3.3, to evaluate "Const." before substituting it into (3.43) (or (3.50)) for $u > 1$ and (3.30) (or (3.49)) for $0 \leq u < 1$.

We can thus get eight particular dimensionless GVF solutions with specified boundary conditions, as shown in (3.62) through (3.69).

(1) For M1 profile ($1 < u < \infty$)

$$x_* = x_{*M1} + u\left[1 - \frac{u^{-N}}{N-1}g\left(\frac{N-1}{N}, u^{-N}\right)\right] - u_{M1}\left[1 - \frac{u_{M1}^{-N}}{N-1}g\left(\frac{N-1}{N}, u_{M1}^{-N}\right)\right] + \frac{\lambda^M}{M-1} \times \left[u^{-M+1}\, g\left(\frac{M-1}{N}, u^{-N}\right) - u_{M1}^{-M+1}\, g\left(\frac{M-1}{N}, u_{M1}^{-N}\right)\right]. \tag{3.62}$$

(2) For M2 profile ($\lambda \leq u < 1$)

$$x_* = x_{*M2} + u\left[1 - g\left(\frac{1}{N}, u^N\right)\right] - u_{M2}\left[1 - g\left(\frac{1}{N}, u_{M2}^N\right)\right] + \frac{\lambda^M}{N-M+1}\left[u^{N-M+1}\, g\left(\frac{N-M+1}{N}, u^N\right) - u_{M2}^{N-M+1}\, g\left(\frac{N-M+1}{N}, u_{M2}^N\right)\right]. \tag{3.63}$$

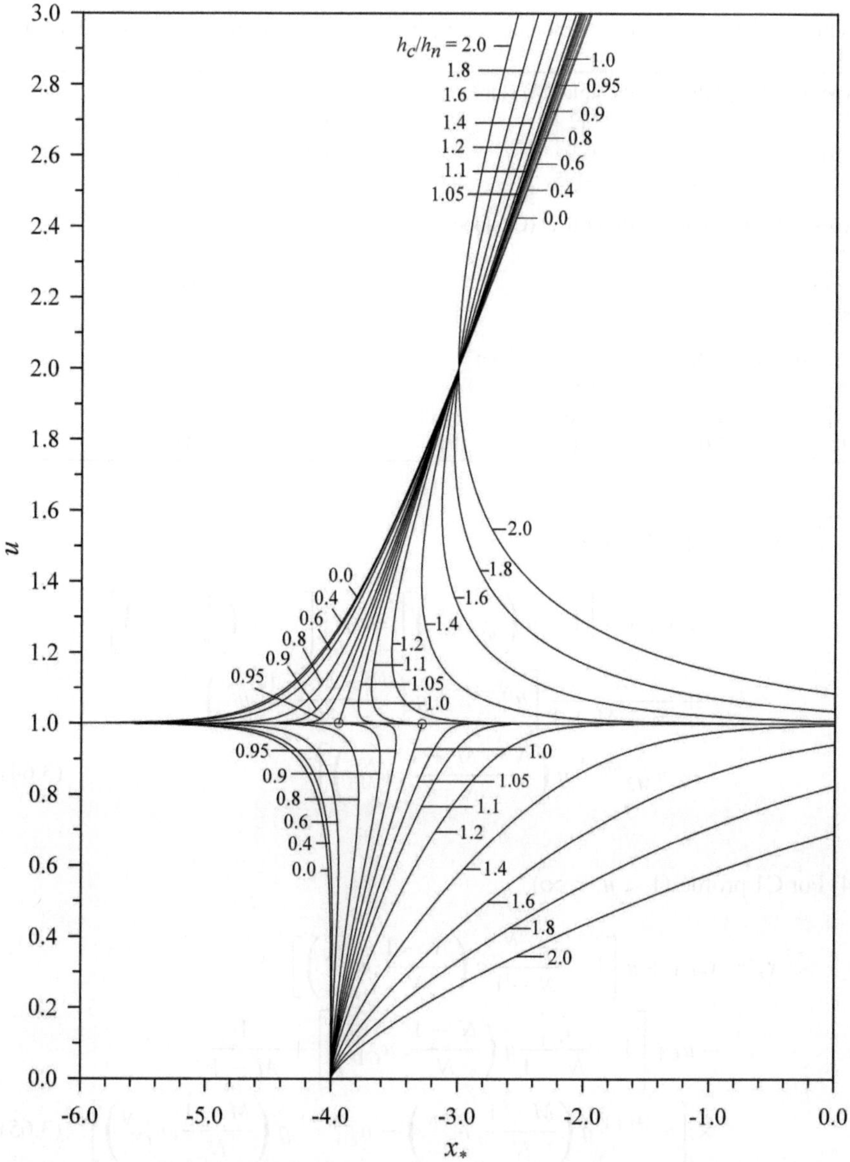

Fig. 3.1 The h_n-based dimensionless GVF profiles for $M = 3$ and $N = 10/3$ plotted against the various values of h_c/h_n. For discerning the profiles from one another with h_c/h_n as a parameter, the boundary conditions of all the profiles plotted for (3.30) and (3.50) for $u > 1$ as well as plotted for (3.43) amd (3.49) for $0 < u < 1$ are arbitrarily set at $(x_*, u) = (-3, 2)$ and $(-4, 0)$, respectively. The theoretical range of h_c/h_n is $0 \leq h_c/h_n < \infty$, but all the profiles are only plotted for $0 \leq h_c/h_n < 2$ herein. Except for $h_c/h_n = 0$, there are two points on each profile to which the slope of the profile id infinite, one at $u = 0$ and the other at $u = h_c/h_n$. *Two small open circles* shown in the graph at $(x_*, u) = (-3.94503, 1)$ and $(-3.28234, 1)$ stand for two singularities on the slopes of the C1 and C3 profiles at $u = 1$, respectively. (Jan and Chen (2012)), ©J. Hydrology, Elsevier B.V., reproduced with permission

Table 3.3 Specified boundary conditions for h_n-based GVF profiles plotted in Figs. 3.2 through 3.4

Types of GVF profiles on the mild (M) slope					
M1		M2		M3	
x_{*M1}	u_{M1}	x_{*M2}	u_{M2}	x_{*M3}	u_{M3}
0.0	3.0	−0.5	0.6	−1.0	0.0
Types of GVF profiles on the critical (C) slope					
C1				C3	
x_{*C1}	u_{C1}			x_{*C3}	u_{C3}
0.0	3.0			−1.8	0.0
Types of GVF profiles on the mild (S) slope					
S1		S2		S3	
x_{*S1}	u_{S1}	x_{*S2}	u_{S2}	x_{*S3}	u_{S3}
0.0	3.0	−1.0	1.8	−2.0	0.0

(3) For M3 profile ($0 \le u \le \lambda$)

$$\begin{aligned} x_* = x_{*M3} &+ u\left[1 - g\left(\frac{1}{N}, u^N\right)\right] - u_{M3}\left[1 - g\left(\frac{1}{N}, u_{M3}^N\right)\right] \\ &+ \frac{\lambda^M}{N-M+1}\left[u^{N-M+1}\, g\left(\frac{N-M+1}{N}, u^N\right)\right. \\ &\left. - u_{M3}^{N-M+1}\, g\left(\frac{N-M+1}{N}, u_{M3}^N\right)\right]. \end{aligned} \tag{3.64}$$

(4) For C1 profile ($1 < u < \infty$)

$$\begin{aligned} x_* = x_{*C1} &+ u\left[1 - \frac{u^{-N}}{N-1} g\left(\frac{N-1}{N}, u^{-N}\right)\right] \\ &- u_{C1}\left[1 - \frac{u_{C1}^{-N}}{N-1} g\left(\frac{N-1}{N}, u_{C1}^{-N}\right)\right] + \frac{1}{M-1} \\ &\times \left[u^{-M+1}\, g\left(\frac{M-1}{N}, u^{-N}\right) - u_{C1}^{-M+1}\, g\left(\frac{M-1}{N}, u_{C1}^{-N}\right)\right]. \end{aligned} \tag{3.65}$$

(5) For C3 profile ($0 \le u < 1$)

$$\begin{aligned} x_* = x_{*C3} &+ u\left[1 - g\left(\frac{1}{N}, u^N\right)\right] - u_{C3}\left[1 - g\left(\frac{1}{N}, u_{C3}^N\right)\right] \\ &+ \frac{1}{N-M+1}\left[u^{N-M+1}\, g\left(\frac{N-M+1}{N}, u^N\right)\right. \end{aligned}$$

$$- u_{C3}^{N-M+1} g\left(\frac{N-M+1}{N}, u_{C3}^{N}\right)\Bigg]. \tag{3.66}$$

(6) For S1 profile ($\lambda \leq u < \infty$)

$$\begin{aligned} x_* = x_{*S1} &+ u\left[1 - \frac{u^{-N}}{N-1} g\left(\frac{N-1}{N}, u^{-N}\right)\right] \\ &- u_{S1}\left[1 - \frac{u_{S1}^{-N}}{N-1} g\left(\frac{N-1}{N}, u_{S1}^{-N}\right)\right] + \frac{\lambda^M}{M-1} \\ &\times \left[u^{-M+1} g\left(\frac{M-1}{N}, u^{-N}\right) - u_{S1}^{-M+1} g\left(\frac{M-1}{N}, u_{S1}^{-N}\right)\right]. \end{aligned} \tag{3.67}$$

(7) For S2 profile ($1 < u \leq \lambda$)

$$\begin{aligned} x_* = x_{*S2} &+ u\left[1 - \frac{u^{-N}}{N-1} g\left(\frac{N-1}{N}, u^{-N}\right)\right] \\ &- u_{S2}\left[1 - \frac{u_{S2}^{-N}}{N-1} g\left(\frac{N-1}{N}, u_{S2}^{-N}\right)\right] + \frac{\lambda^M}{M-1} \\ &\times \left[u^{-M+1} g\left(\frac{M-1}{N}, u^{-N}\right) - u_{S2}^{-M+1} g\left(\frac{M-1}{N}, u_{S2}^{-N}\right)\right]. \end{aligned} \tag{3.68}$$

(8) For S3 profile ($0 \leq u < 1$)

$$\begin{aligned} x_* = x_{*S3} &+ u\left[1 - g\left(\frac{1}{N}, u^{N}\right)\right] - u_{S3}\left[1 - g\left(\frac{1}{N}, u_{S3}^{N}\right)\right] \\ &+ \frac{\lambda^M}{N-M+1}\left[u^{N-M+1} g\left(\frac{N-M+1}{N}, u^{N}\right)\right. \\ &\left. - u_{S3}^{N-M+1} g\left(\frac{N-M+1}{N}, u_{S3}^{N}\right)\right], \end{aligned} \tag{3.69}$$

where (x_{*M1}, u_{M1}), (x_{*M2}, u_{M2}), (x_{*M3}, u_{M3}), (x_{*C1}, u_{C1}), (x_{*C3}, u_{C3}), (x_{*S1}, u_{S1}), (x_{*S2}, u_{S2}) and (x_{*S3}, u_{S3}) are specified boundary conditions for M1, M2, M3, C1, C3, S1, S2 and S3 profiles, respectively.

3.5.5 Examples of GHF-Based GVF Prolfiles with $M = 3$ and $N = 10/3$, Specified λ and BCs

Based on (3.62)–(3.69) and under the conditions $M = 3$ and $N = 10/3$, we can get eight particular GHF-based dimensionless GVF profiles under specified λ-value and Boundary conditions (BCs), as shown in (3.70)–(3.77). These eight particular GHF-based profiles are part of dimensionless GVF profiles that had been plotted in Fig. 3.1.

(1) For M1 profile ($1 < u < \infty$) with $\lambda = 0.6$ and $(x_{*M1}, u_{M1}) = (-3, 2)$:

$$\begin{aligned} x_* = -3 + \Bigg\{ & u\left[1 - \frac{3u^{-10/3}}{7} g\left(\frac{7}{10}, u^{-10/3}\right)\right] \\ & - 2\left[1 - \frac{3 \times 2^{-10/3}}{7} g\left(\frac{7}{10}, 2^{-10/3}\right)\right]\Bigg\} \\ & + \frac{0.6^3}{2}\left[u^{-2} g\left(\frac{3}{5}, u^{-10/3}\right) - 2^{-2} g\left(\frac{3}{5}, 2^{-10/3}\right)\right]. \end{aligned} \tag{3.70}$$

(2) For M2 profile ($\lambda \le u < 1$) with $\lambda = 0.6$ and $(x_{*M2}, u_{M2}) = (-4, 0)$:

$$x_* = -4 + u\left[1 - g\left(\tfrac{3}{10}, u^{10/3}\right)\right] + \tfrac{3 \times 0.6^3}{4} u^{4/3} g\left(\tfrac{2}{5}, u^{10/3}\right). \tag{3.71}$$

(3) For M3 profile ($0 \le u \le \lambda$) with $\lambda = 0.8$ and $(x_{*M3}, u_{M3}) = (-4, 0)$:

$$x_* = -4 + u\left[1 - g\left(\tfrac{3}{10}, u^{10/3}\right)\right] + \tfrac{3 \times 0.8^3}{4} u^{4/3} g\left(\tfrac{2}{5}, u^{10/3}\right). \tag{3.72}$$

(4) For C1 profile ($1 < u < \infty$) with $\lambda = 1$ and $(x_{*C1}, u_{C1}) = (-3, 2)$:

$$\begin{aligned} x_* = -3 + \Bigg\{ & u\left[1 - \frac{3u^{-10/3}}{7} g\left(\frac{7}{10}, u^{-10/3}\right)\right] \\ & - 2\left[1 - \frac{3 \times 2^{-10/3}}{7} g\left(\frac{7}{10}, 2^{-10/3}\right)\right]\Bigg\} \\ & + \frac{1}{2}\left[u^{-2} g\left(\frac{3}{5}, u^{-10/3}\right) - 2^{-2} g\left(\frac{3}{5}, 2^{-10/3}\right)\right]. \end{aligned} \tag{3.73}$$

(5) For C3 profile ($0 \le u < 1$) with $\lambda = 1$ and $(x_{*C3}, u_{C3}) = (-4, 0)$:

$$x_* = -4 + u\left[1 - g\left(\frac{3}{10}, u^{10/3}\right)\right] + \frac{3}{4} u^{4/3} g\left(\frac{2}{5}, u^{10/3}\right). \tag{3.74}$$

(6) For S1 profile ($\lambda \le u < \infty$) with $\lambda = 1.8$ and $(x_{*S1}, u_{S1}) = (-3, 2)$:

$$x_* = -3 + \left\{ u \left[1 - \frac{3u^{-10/3}}{7} g\left(\frac{7}{10}, u^{-10/3}\right) \right] - 2 \left[1 - \frac{3 \times 2^{-10/3}}{7} g\left(\frac{7}{10}, 2^{-10/3}\right) \right] \right\} + \frac{1.8^3}{2} \left[u^{-2} g\left(\frac{3}{5}, u^{-10/3}\right) - 2^{-2} g\left(\frac{3}{5}, 2^{-10/3}\right) \right]. \quad (3.75)$$

(7) For S2 profile ($1 < u \le \lambda$) with $\lambda = 1.2$ and $(x_{*S2}, u_{S2}) = (-3, 2)$:

$$x_* = -3 + \left\{ u \left[1 - \frac{3u^{-10/3}}{7} g\left(\frac{7}{10}, u^{-10/3}\right) \right] - 2 \left[1 - \frac{3 \times 2^{-10/3}}{7} g\left(\frac{7}{10}, 2^{-10/3}\right) \right] \right\} + \frac{1.2^3}{2} \left[u^{-2} g\left(\frac{3}{5}, u^{-10/3}\right) - 2^{-2} g\left(\frac{3}{5}, 2^{-10/3}\right) \right]. \quad (3.76)$$

(8) For S3 profile ($0 \le u < 1$) with $\lambda = 1.8$ and $(x_{*S3}, u_{S3}) = (-4, 0)$:

$$x_* = -4 + u \left[1 - g\left(\frac{3}{10}, u^{10/3}\right) \right] + \frac{3 \times 1.8^3}{4} u^{4/3} g\left(\frac{2}{5}, u^{10/3}\right). \quad (3.77)$$

For more illustration, we substitute $M = 3$ for wide channels and five different N-values, ($N = 3$, $10/3$, $17/5$, $7/2$ and $11/3$) for fully rough flows, one at a time, into each of the eight particular solutions as listed from (3.62) to (3.69), thereby plotting five h_n-based dimensionless GVF profiles on a mild slope ($h_c/h_n = 0.6$), on a critical slope ($h_c/h_n = 1$), and on a steep slope ($h_c/h_n = 1.8$), as depicted in Figs. 3.2, 3.3, and 3.4, respectively, under the specified boundary conditions as indicated in Table 3.3. A comparison of the five dimensionless GVF profiles so plotted in each zone on each type of the three channel slopes with one another may shed light on the effects of h_c/h_n and the N-value on the characteristics of GVF profiles, such as their curvatures and inflection points.

3.6 Validation of the GHF-Based Solutions

We have proved that (3.30) for $0 \le u < 1$ combined with (3.43) for $u > 1$, can represent the complete GHF-based solution of (3.5) for $0 \le u < \infty$, except for $u = 1$. Equations (3.30) and (3.43) on substitution of appropriately prescribed

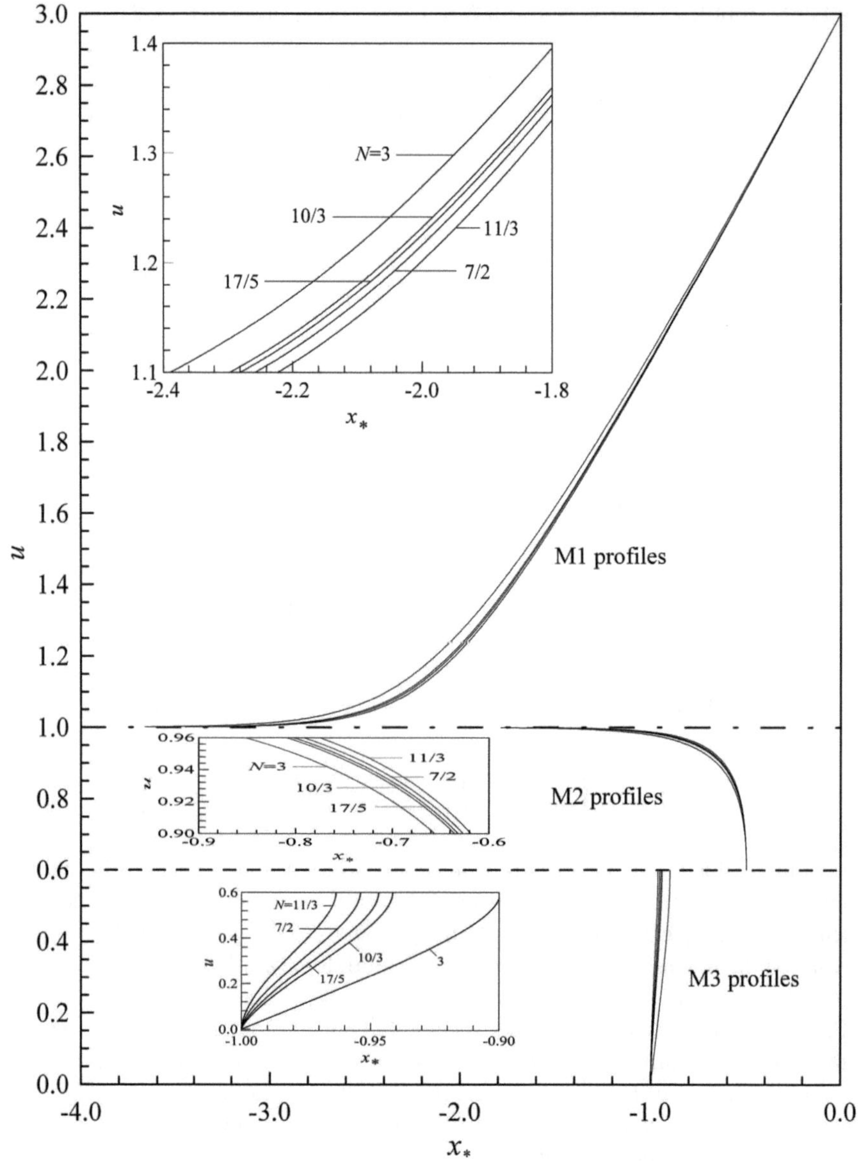

Fig. 3.2 The h_n-based dimensionless GVF profiles on a mild slope ($h_c/h_n = 0.6$) for flow in a wide channel ($M = 3$). The M profiles in each of the *three zones* (M1, M2, and M3) are plotted against five N-values. (Jan and Chen (2012)), ©J. Hydrology, Elsevier B.V., reproduced with permission

boundary conditions will yield eight particular GHF-based GVF solutions, as shown in (3.62)–(3.69). We have proceeded to plot each solution for the five listed N-values in Table 3.2, one at a time, as shown in Figs. 3.2, 3.3 and 3.4. As mentioned earlier, we

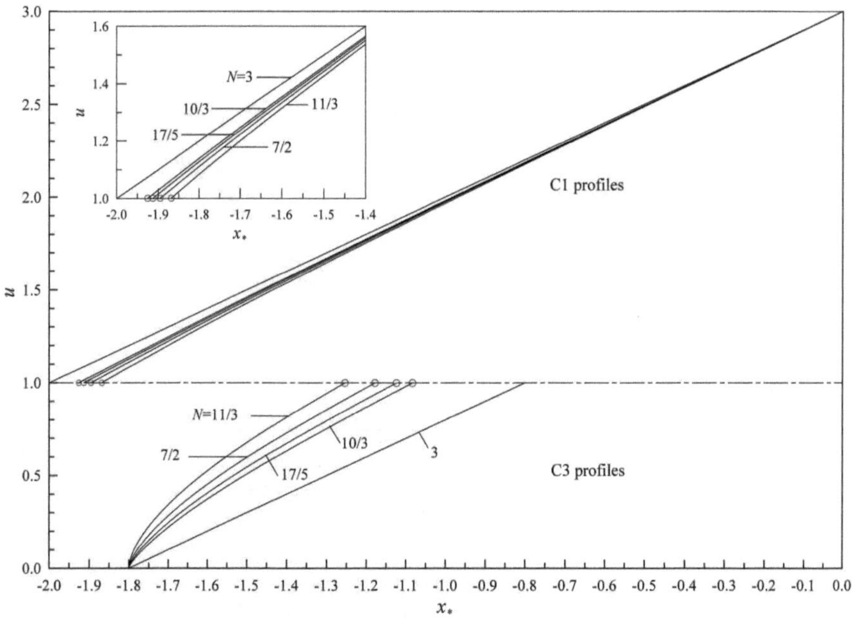

Fig. 3.3 The h_n-based dimensionless GVF profiles on a critical slope ($h_c/h_n = 1$) for flow in a wide channel ($M = 3$). The C profiles in each of the *two zones* (C1 and C3) are plotted against five N-values. A number of *small open circles* plotted in the graph stand for so *many holes* on the slopes of all the C1 and C3 profiles at $u = 1$, connoting that the plotted C profiles for all the N-values except $N = 3$ are undefined at $u = 1$ exactly. (Jan and Chen (2012)), ©J. Hydrology, Elsevier B.V., reproduced with permission

can also use the other analytical solutions based on the ETF to validate the particular GHF-based solutions. Since the two methods of integration in the evaluation of the two integrals of (3.11), (3.12) or (3.13) are exact, both GHF-based and ETF-based solutions so obtained should be exactly identical if there were an analytical way to prove their identity. Unfortunately, we have not yet been able to prove analytically the identity of both solutions for any real values of M and N, except for $M = N = 3$. The ETF-based solution for $M = N = 3$ has often been called the Bresse solution. The analytical poof of the identity between the GHF-based solutions for $M = N = 3$ and the Bresse solution is presented in Appendix E.

It merits mentioning that several advantages exist in using the GHF-based solutions over the alternative solutions expressed in terms of the ETF. Among the most advantageous properties of the GHF-based solutions are recapped below:

1. The parameters, M and N, of the GHF in the GHF-based solutions, (3.30) and (3.43), can be any real numbers, whereas the workable values of M and N used in solving (3.5) via the Mathematica aoftware (or the Maple software) based on the ETF are at most any rational numbers.

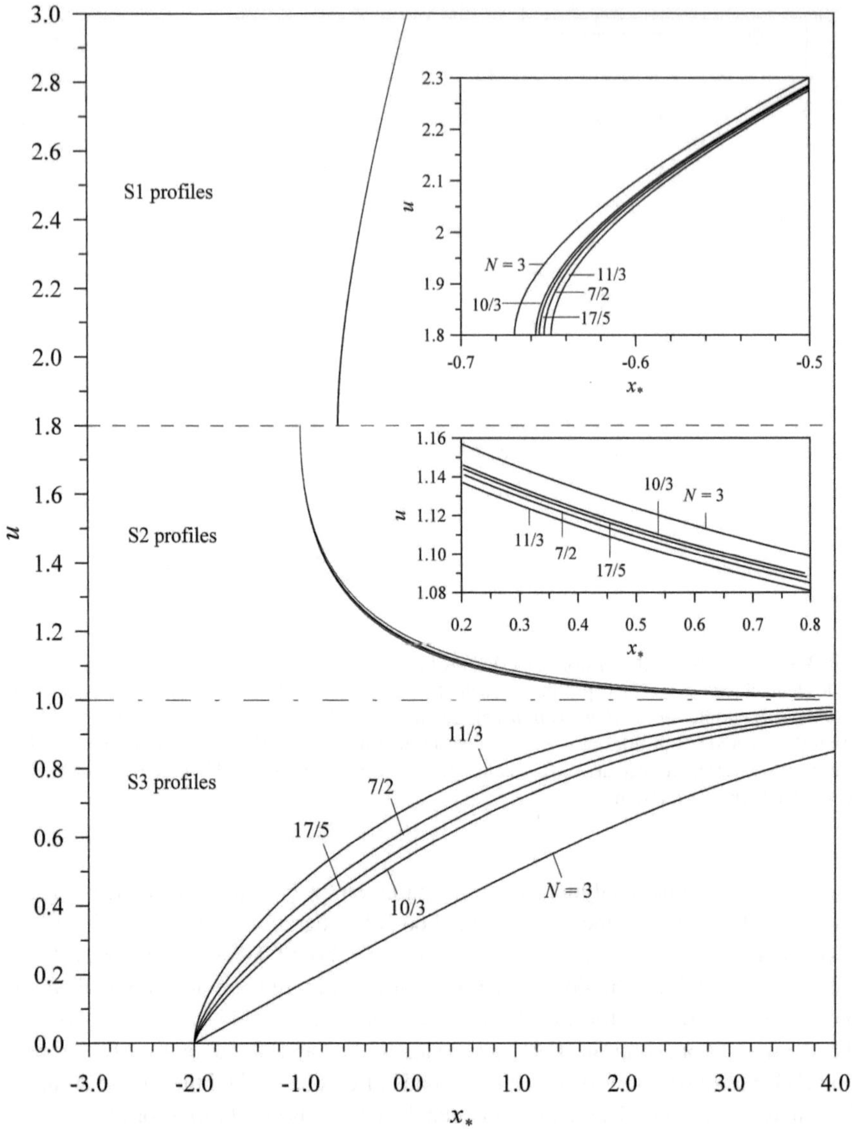

Fig. 3.4 The h_n-based dimensionless GVF profiles on a steep slope ($h_c/h_n = 1.8$) for flow in a wide channel ($M = 3$). The S profiles in each of the *three zones* (S1, S2, and S3) are plotted against five $N-$values. (Jan and Chen (2012)), ©J. Hydrology, Elsevier B.V., reproduced with permission

2. The GHF-based solutions, (3.30) and (3.43), expressed in terms of the orderly behaving parameters, M and N, of the GHF, contrasts distinctly with the lengthy, disorganized alternative analytical solutions based on the ETF, as listed in Table 2.1 for the special case of $M = 3$ and $N = 10/3$, or listed in Appendix D for the ETF-solutions having different N-values.

The above advantages of the GHF should help one to develop a viable procedure to use the GHF-based solutions in the computation of GVF profiles in channels with any regular cross-sectional shape other than wide rectangle, where the values of M and N in the GHF may vary with the flow depth, h.

3.6.1 Solving Equation (3.5) by Use of the ETF

As mentioned earlier, Equations (3.11), (3.12) or (3.13) expressed in an integral form can be used as solutions of (3.5). In this book, we have opted to use a combination of (3.30) and (3.43) as typical GHF-based solutions of (3.5) for $0 \leq u < \infty$, except $u = 1$. The alternative analytical solutions based on the ETF can be obtained through the partial-fraction expansion from the first and second integrals appearing in (3.12) for $M = 3$ and one of the given rational N-values listed in Table 3.2. The ETF-based solutions of the first and second integrals found from the Mathematica software are listed in Appendix D. In this Appendix, equation numbers for the ETF-based solutions of the first integral appearing in (3.12) for the five N-values ($N = 3$, $10/3$, $17/3$, $7/2$ and $11/3$) are numbered, consecutively, from (D.1) through (D.5). An equation for the ETF-based solutions of the second integral appearing in (3.12) for $M = N = 3$ is identical to (D.1) because the expressions of the first and second integrals appearing in (3.12) are identical if $M = N = 3$. The remaining equations listed in Appendix D, i.e., (D.6) through (D.9), represent the ETF-based solutions of the second integral for $(M, N) = (3, 10/3)$, $(3, 17/3)$, $(3, 7/2)$ and $(3, 11/3)$, respectively.

In the simplest case of $M = N = 3$, substitution of the ETF-based solutions, (D.1), into the first and second integrals of (3.12) yields

$$x_* = u - \left[1 - \left(\frac{h_c}{h_n}\right)^3\right] \times \left[\frac{1}{6}\ln\left(\frac{(u^2+u+1)}{(u-1)^2}\right) + \frac{1}{\sqrt{3}}\arctan\left(\frac{(2u+1)}{\sqrt{3}}\right)\right] + \text{Const.}. \quad (3.78)$$

Equation (3.78), often referred to as the Bresse's solution (Bresse (1860) and Chow (1959)), is valid regardless of whether u is in the domain of $0 \leq u < 1$ or $u > 1$ (or $0 \leq u < \infty$, except at $u = 1$, at which the water surface becomes parallel with the channel bed, thus resulting in $x_* = \pm\infty$). Analogous to the GHF-based solutions for $M = N = 3$, (3.78) is also valid for describing the eight types of the h_n-based GVF profiles, namely three on the M slope ($h_c/h_n < 1$), two on the C slope ($h_c/h_n = 1$), and three on the S slope ($h_c/h_n > 1$). In particular, if (3.78) is applied to the C slope, it reduces to (3.51), i.e., a continued straight line without creating a singularity at $u = 1$, a result identical to that obtained from the GHF-based solutions, (3.49) and (3.50).

Another example meriting mention in connection with the alternative analytical solutions expressed in terms of the ETF is the work of Venutelli (2004), who rearranged the equations (D.2) and (D.6) as the ETF-based solutions of the first and second indefinite integrals appearing in (3.12) for $M = 3$ and $N = 10/3$, respectively. The ETF-based GVF solution for $M = 3$ and $N = 10/3$ is given as in Table 2.1.

3.6.2 Comparison of the GHF-Based and ETF-Based Solutions

We have shown that there exist two classes of the analytical solutions of (3.5): One class is the GHF-based solutions, i.e., (3.30) for $0 \le u < 1$ and (3.43) for $u > 1$ with M and N (which can be any real numbers) as parameters. The other is the alternative analytical solutions expressed in terms of the ETF, which are only obtainable for any rational (but not arbitrary real) values of M and N, such as one of the five N-values listed in Table 3.2. Many elementary transcendental functions (ETF) can be expressed by means of Gaussian hypergeometric functions (GHF); therefore, one may deem the GHF as a special transcendental function in the context of the ETF (Whittaker and Watson (1992)). Since both analytical solutions of (3.5) exist, a question may arise as to their uniqueness. Strictly speaking, if both solutions were unique, they should have been different, but they are actually identical. In fact, in the simplest case of $M = N = 3$, we can analytically prove that they are exactly identical, as shown in Appendix E. However, in a general case of $M = 3$ with one of the five N-values listed in Table 3.2 except $N = 3$, it is unlikely that one can analytically prove the identity of both solutions with the present knowledge in mathematics. Instead, we can establish their identity by numerical means, as shown in Tables 3.4, 3.5, and 3.6 for GVF in mild, critical, and steep channels, respectively. In the following, we elaborate on how to verify the identity of both classes of the analytical solutions for a combination of $M = 3$ and each of the five N-values studied.

In Appendix E, we can analytically prove that the complete GHF-based solutions for $M = N = 3$, i.e., a combination of (3.30) for $0 \le u < 1$ and (3.43) for $u > 1$, is exactly identical to the so-called Bresse solution, i.e., (3.78). With the help of the recurrence formula established in Appendix B as well as two appropriately prescribed boundary conditions, one in each domain of u, namely $0 \le u < 1$ or $u > 1$, our proof starts with the formulation of two identities between the two GHF, which consist of three uniquely related function parameters, and a portion of the Bresse solution expressed in logarithmic and inverse trigonometric functions, as shown in (E.10) and (E.11), respectively.

It is much easier to verify numerically than analytically the identity of both GHF-based and ETF-based solutions. Our numerical proof may proceed as follows: Using both solutions, one at a time, we plot five h_n-based dimensionless GVF profiles corresponding to the five N-values given in Table 3.2 on a mild slope ($h_c/h_n = 0.6$), on a critical slope ($h_c/h_n = 1$), and on a steep slope ($h_c/h_n = 1.8$), as illustrated in Figs. 3.2, 3.3, and 3.4, respectively. Upon plotting of such profiles, the dimensionless coordinates corresponding to each of such plotted profiles at two arbitrarily selected dimensionless depths computed from the GHF-based solutions

Table 3.4 Comparison between the two analytical solutions of the GVF equation obtained from the Mathematica software using the ETF and the corresponding GHF for GVF profiles on a mild slope ($h_c/h_n = 0.6$) in wide channels ($M = 3$)

Five N-values imbedded in the GVF equation	M1				M2				M3			
(Given u to calculate x_*)	u	x_*	u	x_*	u	x_*	u	x_*	u	x_*	u	x_*
$N = 3$												
ETF-based solution (x_{*1})	1.001	−3.6668	1.666	−1.4452	0.660	−0.5028	0.840	−0.5780	0.040	−1.0918	0.320	−1.0334
GHF-based solution (x_{*2})	1.001	−3.6668	1.666	−1.4452	0.660	−0.5028	0.840	−0.5780	0.040	−1.0918	0.320	−1.0334
Difference $\mid x_{*1} - x_{*2} \mid$		1.00E−15		1.00E−15		1.00E−15		1.00E−15		1.00E−14		1.00E−15
$N = 10/3$												
ETF-based solution (x_{*1})	1.001	−3.4499	1.666	−1.4121	0.660	−0.5023	0.840	−0.5649	0.040	−1.0563	0.320	−1.0245
GHF-based solution(x_{*2})	1.001	−3.4499	1.666	−1.4121	0.660	−0.5023	0.840	−0.5649	0.040	−1.0563	0.320	−1.0245
Difference $\mid x_{*1} - x_{*2} \mid$		2.98E−14		1.00E−15		1.00E−15		1.00E−15		1.00E−15		1.00E−15
$N = 17/5$												
ETF-based solution (x_{*1})	1.001	−3.4121	1.666	−1.4066	0.660	−0.5022	0.840	−0.5630	0.040	−1.0513	0.320	−1.0231
GHF-based solution(x_{*2})	1.001	−3.4121	1.666	−1.4066	0.660	−0.5022	0.840	−0.5630	0.040	−1.0513	0.320	−1.0231
Difference $\mid x_{*1} - x_{*2} \mid$		1.00E−13		1.00E−15		1.00E−15		1.00E−15		1.00E−15		1.00E−15
$N = 7/2$												
ETF-based solution (x_{*1})	1.001	−3.3583	1.666	−1.3990	0.660	−0.5021	0.840	−0.5603	0.040	−1.0448	0.320	−1.0210
GHF-based solution (x_{*2})	1.001	−3.3583	1.666	−1.3990	0.660	−0.5021	0.840	−0.5603	0.040	−1.0448	0.320	−1.0210
Difference $\mid x_{*1} - x_{*2} \mid$		2.00E−14		1.00E−15		1.00E−15		1.00E−15		1.00E−15		1.00E−15
$N = 11/3$												
ETF-based solution (x_{*1})	1.001	−3.2758	1.666	−1.3877	0.660	−0.5019	0.840	−0.5561	0.040	−1.0359	0.320	−1.0185
GHF-based solution (x_{*2})	1.001	−3.2758	1.666	−1.3877	0.660	−0.5019	0.840	−0.5561	0.040	−1.0359	0.320	−1.0185
Difference $\mid x_{*1} - x_{*2} \mid$		3.02E−14		1.00E−15		1.00E−15		1.00E−15		1.00E−15		1.00E−15

Table 3.5 Comparison between the two analytical solutions of the GVF equation obtained from the Mathematica software using the ETF and the corresponding GHF for GVF profiles on a critical slope ($h_c/h_n = 1.0$) in wide channels ($M = 3$)

Five N-values imbedded in the GVF equation	C1				C3			
(Given u to calculate x_*)	u	x_*	u	x_*	u	x_*	u	x_*
$N = 3$								
ETF-based solution (x_{*1})	1.001	−1.99900	1.666	−1.33400	0.010	−1.98000	0.550	−1.44000
GHF-based solution (x_{*2})	1.001	−1.99900	1.666	−1.33400	0.010	−1.98000	0.550	−1.44000
Difference $\mid x_{*1} - x_{*2} \mid$		1.00E−15		1.00E−15		1.00E−15		1.00E−15
$N = 10/3$								
ETF-based solution (x_{*1})	1.001	−1.92565	1.666	−1.30400	0.010	−1.70705	0.550	−1.37515
GHF-based solution (x_{*2})	1.001	−1.92565	1.666	−1.30400	0.010	−1.70705	0.550	−1.37515
Difference $\mid x_{*1} - x_{*2} \mid$		1.00E−15		1.00E−15		1.00E−15		1.00E−15
$N = 17/5$								
ETF-based solution (x_{*1})	1.001	−1.91309	1.666	−1.29902	0.010	−1.66713	0.550	−1.36386
GHF-based solution (x_{*2})	1.001	−1.91309	1.666	−1.29902	0.010	−1.66713	0.550	−1.36386
Difference $\mid x_{*1} - x_{*2} \mid$		1.00E−15		1.00E−15		1.00E−15		1.00E−15
$N = 7/2$								
ETF-based solution (x_{*1})	1.001	−1.89536	1.666	−1.29209	0.010	−1.61372	0.550	−1.34781
GHF-based solution (x_{*2})	1.001	−1.89536	1.666	−1.29209	0.010	−1.61372	0.550	−1.34781
Difference $\mid x_{*1} - x_{*2} \mid$		1.00E−15		1.00E−15		1.00E−14		1.00E−15
$N = 11/3$								
ETF-based solution (x_{*1})	1.001	−1.86850	1.666	−1.28182	0.010	−1.53873	1.550	−1.32325
GHF-based solution (x_{*2})	1.001	−1.86850	1.666	−1.28182	0.010	−1.53873	1.550	−1.32325
Difference $\mid x_{*1} - x_{*2} \mid$		1.00E−15		1.00E−15		1.00E−15		1.00E−15

Table 3.6 Comparison between the two alternative solutions of the GVF equation obtained from the Mathematica software using the ETF and the corresponding GHF for GVF profiles on a steep slope ($h_c/h_n = 1.8$) in wide channels ($M = 3$)

Five N-values imbedded in the 1-D GVF equation	S1				S2				S3			
(Given u to calculate x_*)	u	x_*	u	x_*	cu	x_*	u	x_*	u	x_*	u	x_*
$N = 3$												
ETF-based solution (x_{*1})	1.810	−0.6693	2.4	−0.4403	1.100	0.7855	1.450	−0.8123	0.110	−1.33583	0.550	1.3300
GHF-based solution (x_{*2})	1.810	−0.6693	2.4	−0.4403	1.100	0.7855	1.450	−0.8123	0.110	−1.33583	0.550	1.3300
Difference $\mid x_{*1} - x_{*2} \mid$		1.00E−15		1.00E−15		2.00E−15		1.00E−15		1.00E−15		1.00E−15
$N = 10/3$												
ETF-based solution (x_{*1})	1.810	−0.6572	3.380	−0.4337	1.100	0.6577	1.450	−0.8212	0.110	−1.7694	0.550	0.0357
GHF-based solution (x_{*2})	1.810	−0.6572	3.380	−0.4337	1.100	0.6577	1.450	−0.8212	0.110	−1.7694	0.550	0.0357
Difference$\mid x_{*1} - x_{*2} \mid$		1.00E−15		1.00E−15		4.00E−15		2.00E−15		1.00E−15		1.00E−15
$N = 17/5$												
ETF-based solution (x_{*1})	1.810	−0.6552	3.380	−0.4327	1.100	0.6353	1.450	−0.8227	0.110	−1.8104	0.550	−0.319
GHF-based solution (x_{*2})	1.810	−0.6552	3.380	−0.4327	1.100	0.6353	1.450	−0.8227	0.110	−1.8104	0.550	−0.319
Difference $\mid x_{*1} - x_{*2} \mid$		1.00E−15		2.00E−15		7.00E−15		5.00E−15		1.00E−15		3.00E−15
$N = 7/2$												
ETF-based solution (x_{*1})	1.810	−0.6524	3.380	−0.4312	1.100	0.6034	1.450	−0.8249	0.110	−1.8581	0.550	−0.3669
GHF-based solution (x_{*2})	1.810	−0.6524	3.380	−0.4312	1.100	0.6034	1.450	−0.8249	0.110	−1.8581	0.550	−0.3669
Difference$\mid x_{*1} - x_{*2} \mid$		1.00E−15		1.00E−15		3.00E−15		1.00E−15		1.00E−15		1.00E−15
$N = 11/3$												
ETF-based solution (x_{*1})	1.810	−0.6484	3.380	−0.4291	1.100	0.5542	1.450	−0.8283	0.110	−1.9116	0.550	−0.6737
GHF-based solution (x_{*2})	1.810	−0.6484	3.380	−0.4291	1.100	0.5542	1.450	−0.8283	0.110	−1.9116	0.550	−0.6737
Difference$\mid x_{*1} - x_{*2} \mid$		1.00E−15		1.00E−15		1.00E−15		1.00E−15		1.00E−15		1.00E−15

are compared with the counterparts computed from the ETF-based solutions. As shown in Tables 3.4, 3.5 and 3.6, we have found that both coordinates so computed are in almost perfect agreement with each other within the nearly zero tolerance prescribed in the Mathematica software. We can thus conclude that both solutions are essentially identical.

3.7 Properties of the GHF-Based Solutions

Eight particular GHF-based solutions of (3.5), i.e., h_n-based dimensionless profiles on the three classified slopes, as listed in Eqs. (3.62)–(3.69), may be used to analyze the properties of the GHF-based solutions, such as the effects of both h_c/h_n and mth power of the power-law resistance formula (or N-value) on the slope of a flow profile, its inflection points, if any, and curvature. The profiles plotted in Figs. 3.2, 3.3 and 3.4 illustrate that the higher the N-value (implying the rougher the flow and the channel boundary), the larger is the curvature of a flow profile. However, to analyze the effects of both h_c/h_n and N-value on the slope of a flow profile, its inflection points, and the existence of singularities, it would be much clearer if we could gain an insight into the dimensionless GVF equation, (3.5), than the solution thereof, i.e., the eight particular GHF-based solutions. How such properties can be analyzed is elaborated next.

3.7.1 *Slopes of Flow Profiles Varying with h_c/h_n and N*

For analyzing the slope of a flow profile, we use (3.5), which can be rewritten in a more meaningful form as follows:

$$\frac{du}{dx_*} = \frac{u^N - 1}{u^{N-M}\left(u^M - \lambda^M\right)}, \qquad (3.79)$$

which shows $du/dx_* = \infty$ at $u = \lambda\,(= h_c/h_n)$ for all N-values; and so it is at $u = 0$ for all N-values except for $N = 3$. By implication, a plot of du/dx_* against u, as represented by the rational function of u on the right-hand side of (3.79), possesses two vertical asymptotes at $u = 0$ and $u = h_c/h_n$ for all values of h_c/h_n except for $h_c/h_n = 1$. In addition to the two vertical asymptotes, it has a horizontal asymptote at $u = 1$ where $du/dx_* = 0$. For illustration, (3.79) is plotted for flow on two mild slopes ($h_c/h_n = 0.6$ and 0.95), a critical slope ($h_c/h_n = 1$), and a steep slope ($h_c/h_n = 1.8$), as shown in Figs. 3.5 through 3.8, respectively. These figures show that there are singularities on the C profiles and two points of inflection on each of the M profiles. In what follows, we analyze such particular properties of GVF profiles, which merit special attention.

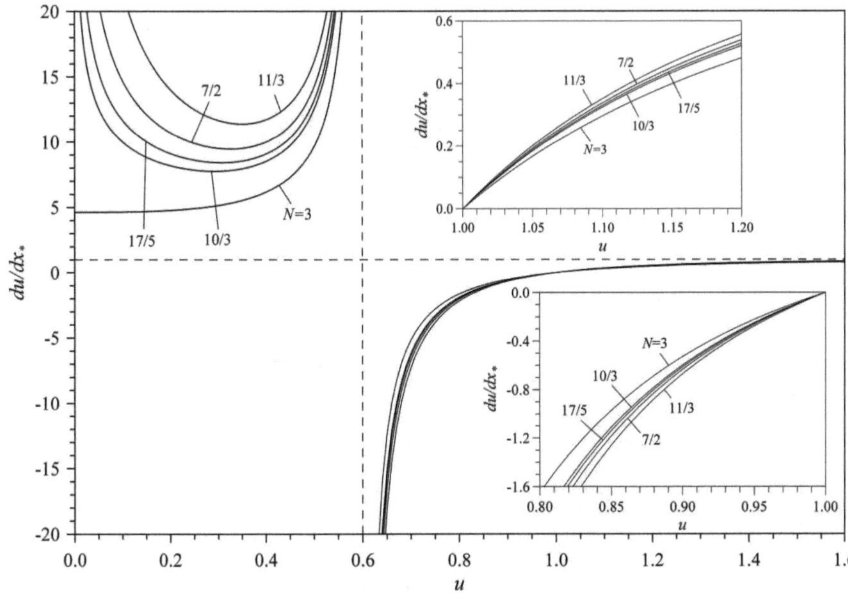

Fig. 3.5 The slope of a gradually-varied flow profile (i.e., du/dx_*) is plotted against u for M profiles with $h_c/h_n = 0.6$, $M = 3$, and various $N-$values. The point of inflection on an M profile takes place at a point where du/dx_* is an extremum (i.e., either a minimum or a maximum). There are supposedly two inflection points on each of all M profiles having various $N-$values except for $N = 3$. However, one inflection point appearing on an M1 profile ($u > 1$) with a maximum value of du/dx_* cannot be shown in the figure for all $N-$values plotted because it stays beyond the maximum $u = 1.6$ imposed on the abscissa of the graph. In contrast, the other inflection point appearing on an M3 profile ($0 \leq u \leq h_c/h_n$) with a minimum value of du/dx_* is clearly shown in the figure for all $N-$values except $N = 3$

3.7.2 Singularities of the Rational Function Representing the Slopes of C1 and C3 Profiles

We can readily prove that except for $M = N = 3$, both denominator and numerator on the right-hand side of (3.79) for $h_c/h_n = 1$ are zero at $u = 1$, connoting a singularity at $u = 1$ in the rational function representing the right-hand side of (3.79). This is equivalent to stipulating that the GHF-based solutions for the C3 and C1 profiles, as described by (3.49) and (3.50), respectively, or their slopes, (3.79), are undefined at $u = 1$. The singularity is sometimes called an "indeterminate expression," which is output from the Mathematica software after substituting $u = 1$ into (3.49) or (3.50).

In Fig. 3.7, small open circles are used to mark the singularities on the plotted slopes of the C profiles at $u = 1$. We have found that the singularities of the rational function representing the right-hand side of (3.79) for $N = 10/3, 17/5, 7/2$ and $11/3$ are approximately located at $(u, du/dx_*) = (1, 1.11111)$, $(1, 1.13333)$,

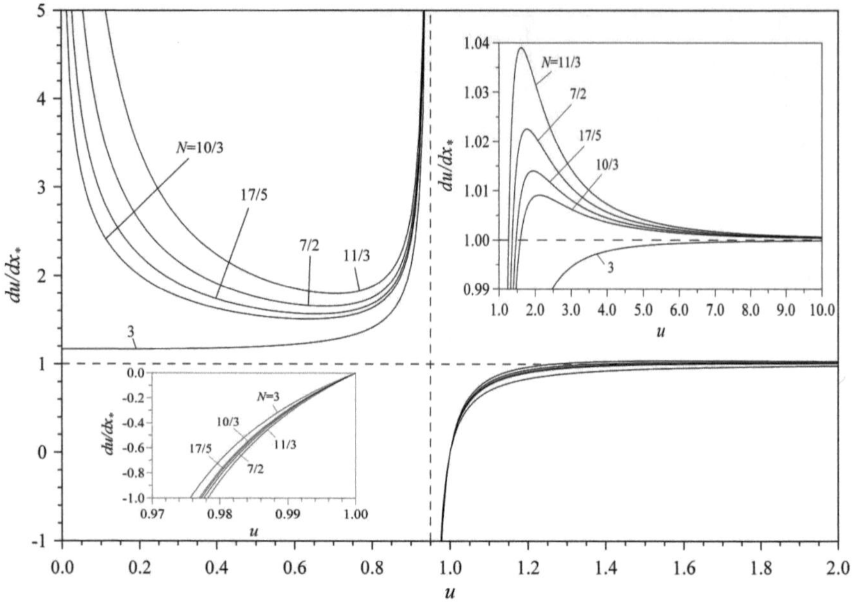

Fig. 3.6 The slope of a gradually-varied flow profile (i.c, du/dx_*) is plotted against u for M profiles with $h_c/h_n = 0.95$, $M = 3$, and various $N-$values. The point of inflection on an M profile takes place at a point where du/dx_* is an extremum (i.e., either a minimum or a maximum). There are two inflection points on each of all M profiles having various $N-$values except for $N = 3$, i.e., one on an M1 profile ($u > 1$) and the other on an M3 profile ($0 \le u \le h_c/h_n$)

(1, 1.16667), and (1, 1.22222), respectively, if the prescribed tolerance of the truncation error is 1.0×10^{-5}. Because (3.79) rather than the GHF-based solutions, i.e., (3.49) and (3.50), is used in analyzing the singularities, the locations of the singularities so determined in the du/dx_* versus u graph is independent of the boundary conditions, thus being unique as far as the computed locations of the singularities are concerned. There is no singularity for $N = 3$ because (3.79) on substitution of $h_c/h_n = 1$ and $M = N = 3$ reduces to $du/dx_* = 1$, i.e., a horizontal asymptote of the du/dx_* versus u plot in Fig. 3.7.

3.7.3 Points of Inflection on the M1 and M3 Profiles

The point of inflection on a flow profile takes place at a point on the du/dx_* versus u plane, where du/dx_* is an extremum (i.e., either a minimum or a maximum). The value of u, at which the inflection point occurs, can be determined from the condition under which $d^2u/dx_*^2 = 0$. We can derive such a condition from (3.5) or (3.79) with the result

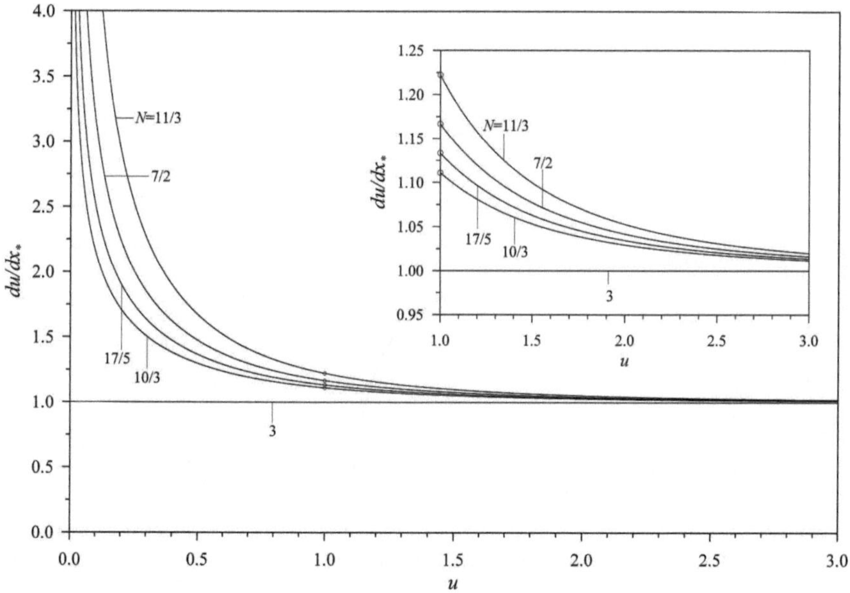

Fig. 3.7 The slope of a gradually-varied flow profile (i.e., du/dx_*) is plotted against u for C profiles with $h_c/h_n = 1$, $M = 3$, and various N-values. There is no extremum in the plotted curves; therefore, no point of inflection exists on C profiles. Small open circles shown in the graph stand for singularities on the slopes of the tangents to C profiles at $u = 1$, connoting that a rational function expressing the slopes of the C profiles is undefined at $u = 1$ exactly. The singularities of the rational function for $N = 10/3, 17/5, 7/2$ and $11/3$ are approximately located at $(u, du/dx_*) =$ (1, 1.11111), (1, 1.13333), (1, 1.16667), and (1, 1.22222), respectively, if the prescribed tolerance of the truncation error is 1.0×10^{-5}

$$\frac{d^2u}{dx_*^2} = -\frac{(u^N - 1)\left[M\lambda^M u^N - Nu^M + (N - M)\lambda^M\right]}{u^{2(N-M)+1}\left(u^M - \lambda^M\right)^3}. \tag{3.80}$$

Equating the right-hand side of the equal sign in (3.80) to zero yields the only possible condition under which $d^2u/dx_*^2 = 0$, i.e.,

$$M\lambda^M u^N - Nu^M + (N - M)\lambda^M = 0. \tag{3.81}$$

It should be noted that Eq. (3.81) becomes invalid for $\lambda\ (= h_c/h_n) = 0$ or $N = M$ because it only yields a trivial solution, $u = 0$. In particular, (3.81) for $M = 3$ and $N = 10/3$ reduces to

$$3\lambda^3 u^{10-3} - \frac{10}{3}u^3 + \frac{1}{3}\lambda^3 = 0, \tag{3.82}$$

which is indeed identical to the equation obtained by Gunder (1943) and Venutelli (2004). Chow (1957) was among the first persons who recognized the coexistence of two inflection points, one on each of M1 and M3 profiles. Venutelli found two physically possible solutions from (3.82) for $h_c/h_n = 0.6847$ as follows: $u = 41.440$

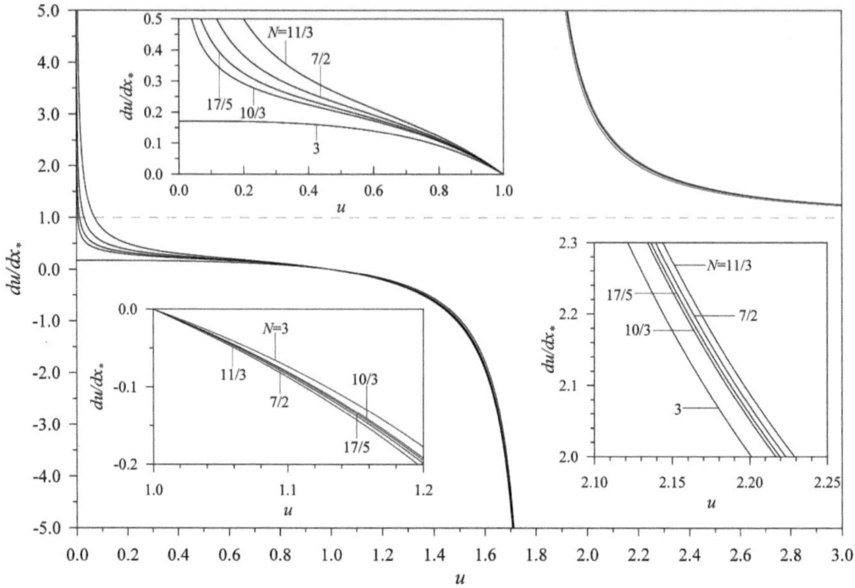

Fig. 3.8 The slope of a gradually-varied flow profile is plotted against u for S profiles with $h_c/h_n = 1.8$, $M = 3$, and various N-values. There is no extremum in the plotted curves; therefore, no point of inflection exists on S profiles

on the M1 profile and $u = 0.343$ on the M3 profile. For validation, the accuracy of Venutelli's solutions is examined by solving (3.82) for u using the Mathematica software. We obtained two inflection points at $u = 41.4733$ and $u = 0.342659$ on M1 and M3 profiles, respectively. The fact that the solutions obtained by us are more precise than those by Venutelli suggests merely that the prescribed tolerance of the truncation error in our solutions is likely smaller than that in Venutelli's solutions.

To show the existence of inflection points on a GVF profile, we plot graphs of du/dx_* against u curves, as shown in Figs. 3.5, 3.6, 3.7 and 3.8. If there is an extremum on any curve, as shown in Figs. 3.5 and 3.6 for all M profiles with $h_c/h_n = 0.6$ and 0.95, respectively, there must be an inflection point. We can easily recognize a minimum on each of M3 profiles ($0 \leq u \leq h_c/h_n$) for all N-values except $N = 3$, but a maximum on each of M1 profiles ($u > 1$) does not seem so easily recognizable for both $h_c/h_n = 0.6$ and 0.95 unless a large scale of du/dx_* on the ordinate is adopted to magnify an infinitesimal rise of du/dx_* versus u curve between $u = 1.6$ and 2.2, which eventually flattens onto the horizontal asymptote at $du/dx_* = 1$ as $u \to \infty$ (see an inset at the upper right corner in Fig. 2.6). In contrast, Figs. 3.7 and 3.8 show no existence of an extremum, i.e., no inflection point on any C and S profiles, respectively.

To locate two points of inflection on each M profile graphically, we set up the left-hand side of the equal sign in (3.81) as a function of u, i.e. $f(u)$, which is plotted against u for $M = 3$ and various N-values, as shown in Figs. 3.9 and 3.10 for $h_c/h_n = 0.6$ and 0.95, respectively. Two inflection points at which $f(u) = 0$ [i.e.,

Table 3.7 The h_n-based dimensionless flow depths at the two inflection points (IPs), One on the M1 profile and the other on the M3 profile, computed from (3.81) for the four N-values of GVF profiles on two mild slopes ($h_c/h_n = 0.6$ and 0.95) in wild channels ($M = 3$)

Hydraulic exponents	$\lambda = h_c/h_n = 0.6$		$\lambda = h_c/h_n = 0.95$		Given the values of λ, M and N to solve equation (3.81)
(M, N)	u at IP on M1 profile $1 < u < \infty$	u at IP on M3 profile $0 \leq u \leq \lambda$	u at IP on M1 profile $1 < u < \infty$	u at IP on M3 profile $0 \leq u \leq \lambda$	$M\lambda^M u^N - N u^M + (N - M)\lambda^M = 0$
(3, 10/3)	136.117	0.2916	3.581	0.6335	$3\lambda^3 u^{10/3} - \frac{10}{3} u^3 + \frac{1}{3}\lambda^3 = 0$
(3, 17/5)	63.061	0.3067	1.9403	0.6532	$3\lambda^3 u^{17/5} - \frac{17}{5} u^3 + \frac{2}{5}\lambda^3 = 0$
(3, 7/2)	29.173	0.3255	1.7708	0.6765	$3\lambda^3 u^{7/2} - \frac{7}{2} u^3 + \frac{1}{2}\lambda^3 = 0$
(3, 11/3)	13.460	0.3505	1.6071	0.7054	$3\lambda^3 u^{11/3} - \frac{11}{3} u^3 + \frac{2}{3}\lambda^3 = 0$

the $f(u)$ versus u curve in each figure or inset intersecting its abscissa] are located on each $f(u)$ versus u curve for all the given N-values except for $N = 3$, one on an M1 profile ($u > 1$) and the other on an M3 profile ($0 \leq u \leq h_c/h_n$).

To study the effects of the N-value on the inflection points, we can numerically calculate the two inflection points for both $h_c/h_n = 0.6$ and 0.95, using (3.81) upon substitution of $M = 3$ and $N = 10/3$, 17/5, 7/2, and 11/3, one at a time, as shown in Table 3.7. It is noted that both inflection points so determined on the M1 and M3 profiles with one of the four N-values studied, as tabulated in Table 3.7, are all expressed in terms of u, irrespective of the boundary conditions. Therefore, the dimensionless longitudinal coordinate, x_*, of each inflection point can be calculated from (3.62) for the M1 profile or (3.64) for the M3 profile on substitution of $M = 3$, the N-value under study, one of the corresponding two boundary conditions, either (x_{*M1}, u_{M1}) or (x_{*M3}, u_{M3}), and the corresponding inflection point (i.e., the u-value just determined).

The numerical solutions of u obtained from (3.81) for various values of h_c/h_n and N reveals the validity of (3.81) in the range of $0 < h_c/h_n < 1$, excluding both end points at $h_c/h_n = 0$ and 1. On the one hand, as $h_c/h_n \to 0$ (i.e., $h_n \to \infty$, thus resulting in the undefined u), one of the two solutions of u obtained from (3.81) decreases in value and approaches zero, while the other solution increases in value and approaches infinity. On the other hand, moving to the other end point of the range as $h_c/h_n \to 1$, both solutions of u obtained from (3.81) approach unity, one from below and the other from above, but $u \neq 1$ at $h_c/h_n = 1$ for all N-values except $N = M$ $(= 3)$, connoting no solution of (3.81). In Fig. 3.11, we plot eight solution curves representing the paths of the two inflection points of the M1 and M3 profiles on the u versus h_c/h_n plane with N as a parameter for the four N-values studied, thereby showing the combined effects of h_c/h_n and N on the locations of the two inflection points.

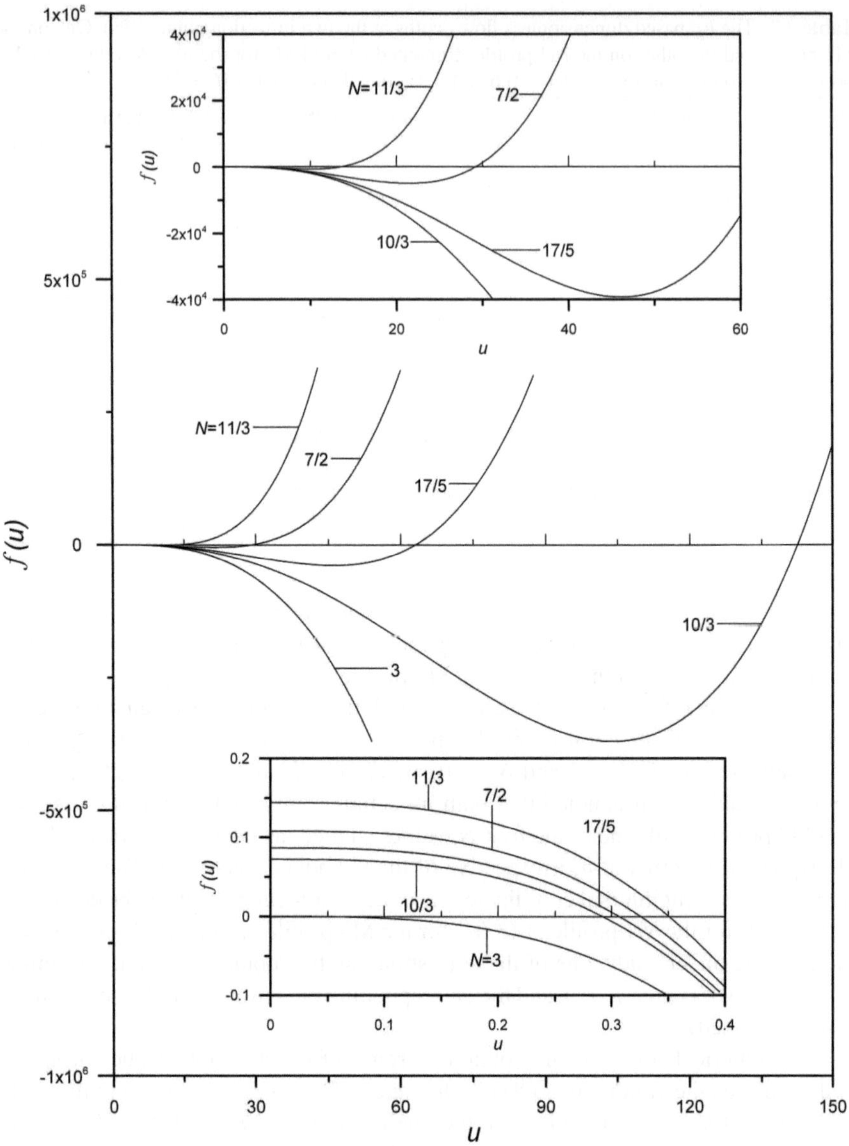

Fig. 3.9 Plot of solution curves, $f(u)$ versus u, for determining the points of inflection on GVF profiles with $h_c/h_n = 0.6$, $M = 3$, and various N-values. Two inflection points at which $f(u) = 0$ can be determined from each solution curve for a given N-value except $N = 3$, one on an M1 profile ($u > 1$) and the other on an M3 profile ($0 \leq u \leq h_c/h_n$)

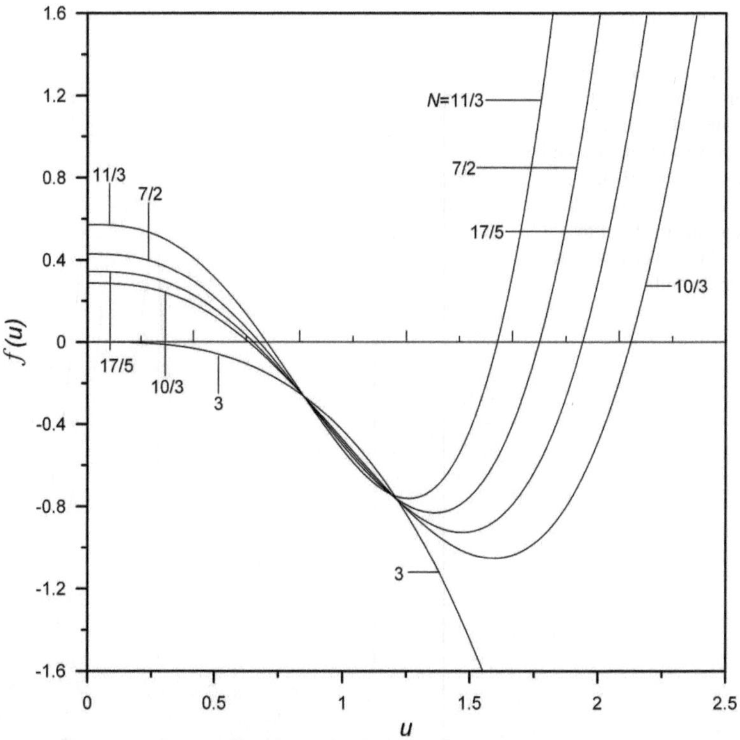

Fig. 3.10 Plot of solution curves, $f(u)$ versus u, for determining the points of inflection for GVF profiles with $h_c/h_n = 0.95$, $M = 3$, and various N-values. Two inflection points at which $f(u) = 0$ can be determined from each solution curve for a given N-value except $N = 3$, one on an M1 profile ($u > 1$) and the other on an M3 profile ($0 \le u \le h_c/h_n$)

3.8 Discussion

3.8.1 Applicability of the GHF-Based Solutions in Perspective

In this book, we treat the hydraulic exponents, M and N, as constants, thereby averting the difficulty in obtaining the GHF-based solutions from (3.5) for the dimensionless GVF profiles in channels with depth-dependent M- and N-values. The mathematical capability of the parameters, M and N, in the GHF to accept any real numbers, which encompass rational numbers, will enable one to expand the GHF-based solutions for GVF profiles in wide channels to those in channels with regular cross-sectional shape, such as rectangles, triangles, trapezoids, and circles, in which the values of M and N vary with the flow depth, h. Such superiority of the GHF-based solutions over the ETF-based solutions, in which the values of M and N can solely accept rational numbers will exclusively enable one to apply the

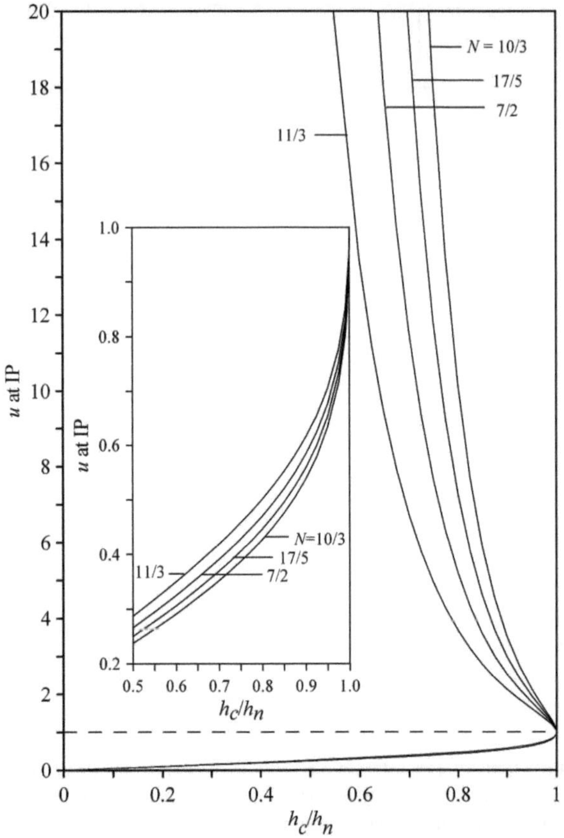

Fig. 3.11 Plot of eight solution curves representing the two values of u at the inflection points (IP) of the M1 and M3 profiles against the various h_c/h_n and N-values. $N = 10/3$, $17/5$, $7/2$, and $11/3$, each corresponding to the mth power of the power-law flow resistance formula

GHF-based solutions to the numerical computation of GVF profiles in channels with cross-sectional shape other than wide rectangles.

It merits attention when the GHF-based solutions so obtained for constant M- and N-values are applied to a general case, in which GVF profiles in channels with cross-sectional shape other than wide rectangles are computed numerically. Success in such general applications depends also on one's ability to evaluate precisely the variations of the M- and N-values with h. Thus, before applying numerically the GHF-based solutions to such a general case, yet to be resolved is an issue on how to determine precisely the variations of the M- and N-values with h in a channel with cross-sectional shape other than a wide rectangle.

For applying the GHF-based solution, (3.30) and (3.43), to the numerical computation of GVF profiles in channels with cross-sectional shape other than wide rectangles, one may imagine a GHF-based solution space, which is composed of a number of the GHF-based solutions corresponding to a number of combinations of the M- and N-values. It may be further imagined that the role of such a GHF-based solution space playing in the numerical computation of GVF profiles in prismatic channels with any cross-sectional shape describable by h-dependent M- and N-values is equivalent to that of the VFF table playing in the interpolation of the VFF-values for use in the Bakhmeteff-Chow procedure to compute such GVF profiles numerically.

The assumption of constant hydraulic exponents is satisfactory in the evaluation the GVF profiles in most rectangular and trapezoidal channels. However, the hydraulic exponents may vary appreciably with respective to the flow depth when the channel section has abrupt changes in geometry of cross section or is topped with a gradually closing crown, such as circular channels. In such cases, the channel length should be divided into a number of reaches in each of which the hydraulic exponents appear to be constant (Chow (1959)), or an improved method or innovative method is needed to overcome this problem (Zaghloul (1998)). The direct-integration method for solving the dimensionless GVF equation using the GHF has been successful developed as shown in the previous sections. However, one should note that the GHF-based GVF solutions only take care of the frictional energy loss without the consideration of the eddy loss. Therefore, the GHF-based solutions obtained here is suitable for GVF in prismatic channels. The energy loss in a natural channel that usually has non-prismatic channel sections is the sum of friction and eddy losses. A suitable numerical method or advance hydrologic model is still needed to compute the GVF profile in a natural channel. The basic methodology used in the computation of GVF profiles in natural channels is the standard-step method and the computations can be carried out using well tried out software, like HEC-RAS and MIKE 21 (Subramanya (2009)).

3.8.2 Role of h_c/h_n in the Domain of the GHF-Based Solution Space

In case for sequential applications of the GHF-based solution to the numerical computation of GVF profiles in a series of channels, the continuity of h_c/h_n in the domain of the GHF-based solution space for the given values of M and N is crucial to success in such computation. One can clearly see from Fig. 3.1 that GVF profiles vary continuously throughout the domain of the GHF-based solution space as h_c/h_n changes from 0 to ∞ though all the profiles are not plotted beyond $h_c/h_n = 2$ and $u = 3$, i.e. a limit imposed on the scale of the ordinate in the graph. In particular, if $h_c/h_n \to 0$, (3.30) and (3.43) reduce respectively to

$$x_* = u\left[1 - g\left(\frac{1}{N}, u^N\right)\right] + \text{Const.} \tag{3.83}$$

and

$$x_* = u\left[1 - \frac{u^{-N}}{N-1} g\left(\frac{N-1}{N}, u^{-N}\right)\right] + \text{Const.}, \tag{3.84}$$

in which "Const." can be evaluated using the two boundary conditions at $(x_*, u) = (-4, 0)$ and $(-3, 2)$, respectively. Consequently, (3.83) and (3.84) on substitution of the respective "Const." so evaluated yields, respectively,

$$x_* = u\left[1 - g\left(\frac{1}{N}, u^N\right)\right] - 4 \tag{3.85}$$

and

$$x_* = u - \frac{u^{-N+1}}{N-1} g\left(\frac{N-1}{N}, u^{-N}\right) + \frac{2^{-N+1}}{N-1} g\left(\frac{N-1}{N}, 2^{-N}\right) - 5. \tag{3.86}$$

Eqs. (3.85) and (3.86) for $N = 10/3$ are plotted in Fig. 3.1 as limits of (3.30) for $0 < u < 1$ and (3.43) for $u > 1$, respectively, for $h_c/h_n \to 0$.

3.8.3 Reclassification of the Critical Profiles and Points of Infinite Profile Slopes

An inspection of the pattern of GVF profiles in Fig. 3.1 shows that the entire domain of the GHF-based solution space for $M = 3$ and $N = 10/3$ can be judiciously grouped into two major regions with two curves, called the critical (or C) profiles marking $h_c/h_n = 1$, being used to separate them. Profiles plotted in the region for $0 < h_c/h_n < 1$ are called the mild (or M) profiles, and those in the region for $h_c/h_n > 1$ the steep (or S) profiles.

According to Chow (1959), each of the M, S, and C profiles so classified can be further divided into three types, but the further division of the C profiles into three types rather than two types cannot be justified in view of the existence of the two singularities at $u = 1$, as marked by two small circles in Fig. 3.1. Before reclassifying the C profiles by virtue of what we have learned in this study, we first specify the three types of the M and S profiles as follows:

1. Three types of the M profiles in the three zones, such as $u > 1$ (i.e., $h > h_n$), $h_c/h_n \le u < 1$ (i.e., $h_c \le h < h_n$), and $0 \le u \le h_c/h_n$ (i.e., $0 \le h \le h_c$), are referred to as M1, M2, and M3, respectively.
2. Three types of the S profiles in the three zones, such as $u \ge h_c/h_n$ (i.e., $h \ge h_c$), $1 < u \le h_c/h_n$ (i.e., $h_n < h \le h_c$), and $0 \le u < 1$ (i.e., $0 \le h < h_n$), are referred to as S1, S2, and S3, respectively.

It seems unjustifiable that the C profiles can be further classified into three types, C1, C2, and C3, corresponding to the three zones in a channel, such as $u > h_c/h_n = 1$ (or $h > h_c = h_n$), $u = 1$ (or $h = h_c = h_n$), and $0 \le u < h_c/h_n = 1$ (or $0 \le h < h_c = h_n$), respectively, because the so-called C2 profile classified in the middle zone ($u = 1$) by Chow (1959) is nothing but a singularity (which is undefined), though a case for $N = 3$ may be excluded. In any event, to deem the uniform flow in case only of $N = 3$ as a type of the C2 profiles is unwarranted, thus excluding C2 from our classification.

Each of the M and S profiles has two points at which the slope of the profile (i.e., du/dx_*) is infinite, namely one point at $u = h_c/h_n$ and the other at $u = 0$, for all N-values except for $N = 3$, as shown in Figs. 3.2, 3.3 and 3.4. In fact, both infinite-slope points on each M profile excluding the one for $h_c/h_n = 0$ fall below $u = 1$, whereas one of such two infinite-slope points on each S profile lies above $u = 1$ and the other below $u = 1$, i.e. at $u = 0$ to be exact. Approaching asymptotically to the limit of GVF profiles for $h_c/h_n = 0$, both infinite-slope points on the M profile collapse at $u = 0$. In contrast to the M and S profiles, there exists only one infinite-slope point at $u = 0$ on the C3 profile, as shown in Fig. 3.3, but all the M, C, and S profiles for $N = 3$ is a special case, in which there is no infinite-slope profile at $u = 0$, as illustrated in Figs. 3.2, 3.3 and 3.4.

Though the effects of the mth power of the power-law flow resistance formula (or N-value) on the GVF profile and its slope are not so appreciably distinct through a simple comparison among the scaled profiles plotted in Figs.3.2, 3.3 and 3.4, suffice it to say that by and large the higher the N-value (implying the rougher the flow and the channel boundary), the greater is the curvature of a flow profile. A further comparison of the plotted slopes of all types of the GVF profiles, i.e. du/dx_* versus u, for all N-values except for $N = 3$, as shown in Figs. 3.5, 3.6, 3.7 and 3.8, reconfirms not only the results found in Figs. 3.2, 3.3 and 3.4, but also the existence of two inflection points on the M profiles. Worth elaborating below are both inflection points.

3.8.4 Identification of Inflection Points on GVF Profiles

One can readily identify the existence of inflection points on any GVF profile by extrema from a plot of du/dx_* versus u, as shown in Figs 3.5, 3.6, 3.7 and 3.8, as mentioned earlier. In general, an extremum lies between two points with the two unique values of du/dx_* : One is a vertical asymptote at $u = 0$ and the other is either another vertical asymptote at $u = h_c/h_n$ or a horizontal asymptote at $du/dx_* = 1$. As shown in Figs. 3.5 and 3.6, the minimum of du/dx_* versus u curve for each M3 profile ($0 \leq u \leq h_c/h_n$) lies between the two vertical asymptotes at $u = 0$ and at $u = h_c/h_n$, whereas a du/dx_* versus u curve for each combined M1-M2 profile ($u \geq h_c/h_n$), which originates at the negative vertical asymptote at $u = h_c/h_n$, rises slightly above the horizontal asymptote ($du/dx_* = 1$) to the maximum before flattening out gradually onto the horizontal asymptote at $du/dx_* = 1$ as $u \to \infty$. This is clearly shown in an inset at the upper right corner of Fig. 3.6 although a tiny rise of du/dx_* versus u curve to the maximum above the horizontal asymptote is negligibly small, especially for $N = 10/3$.

Significant effects of the values of h_c/h_n and N on the locations of the two inflection points are evident from the co-existence of both maxima and minima on du/dx_* versus u curves. The locations of both extrema vary with the values of h_c/h_n and N, as shown in Figs. 3.5 and 3.6. Both inflection points exist for all the N-values studied except for $N = 3$. An analysis of either graphical solutions in Figs. 3.9 and 3.10 or numerical solutions obtained by (3.81), as tabulated in Table 3.7, can disclose a trend of the displacement of the locations of both inflection points. As shown in Fig. 3.11, the closer to unity the value of h_c/h_n and/or the higher the N-value, the closer the two inflection points move toward $u = 1$ as their asymptotic limit. Conversely, in the reverse trend, Fig. 3.11 also shows that the closer to zero the value of h_c/h_n and/or the smaller the N-value, the farther the two inflection points stay apart and eventually go to extremes, i.e., one of them approaching zero and the other going to infinity.

3.8.5 Curvature of GVF Profiles

The curvature, K_u, of a GVF profile at any point (x_*, u) can be expressed from calculus as

$$K_u = \frac{\left|\dfrac{d^2u}{dx_*^2}\right|}{\left[1 + \left(\dfrac{du}{dx_*}\right)^2\right]^{3/2}}, \tag{3.87}$$

which on substitution of d^2u/dx_*^2 from (3.80) and du/dx_* from (3.79) yields

$$K_u = \frac{\left|-(u^N - 1)\left[M\lambda^M u^N - Nu^M + (N - M)\lambda^M\right]\right|}{\mid u^{-N+M+1} \mid \{u^{2(N-M)}(u^M - \lambda^M)^2 + (u^N - 1)^2\}^{3/2}}. \tag{3.88}$$

Evidently, (3.88) shows that K_u is zero at the inflection point $d^2u/dx_*^2 = 0$ by virtue of (3.81) and for the case of $u = 1$. (i.e., $h = h_n$), where the GVF profile becomes parallel with the bed (i.e., uniform flow). Because $N - M > 0$ and $-N + M + 1 > 0$ for all the N-values studied herein, it is true that $du/dx_* = \infty$ and $K_u = \infty$ for all profiles at the channel bed (i.e., $u = 0$) by virtue of (3.79) and (3.88), respectively. Furthermore, (3.79) shows that $du/dx_* = \infty$ at $u = h_c/h_n$, though the value of K_u computed from (3.88) should be relatively large, but not infinity, for every profile, as illustrated also in Fig. 3.1 for $N = 10/3$.

For comparison, it may intrigue one if the curvature at a specific point on the GVF profile is calculated and plotted. For example, for K_u at $u = \lambda$ $(= h_c/h_n)$, (3.88) upon substitution of $u = \lambda$ yields

$$K_u \mid_{u=\lambda} = \left|\frac{-M\lambda^{N-1}}{(\lambda^N - 1)}\right|, \tag{3.89}$$

which indicates that K_u at $u = h_c/h_n$ for the given N-value further depends on the value of h_c/h_n. Obviously, if h_c/h_n is zero, K_u at $u = h_c/h_n$ is zero too. On the

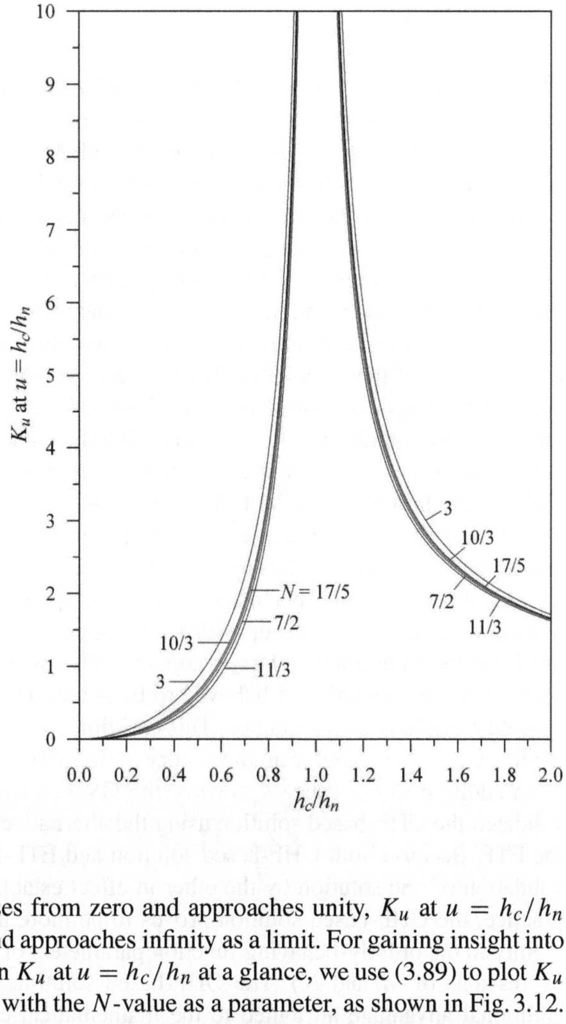

Fig. 3.12 Plot of the curvature, K_u, at $u = h_c/h_n$ against h_c/h_n with the N-value as a parameter for GVF profiles in sustaining channels. The theoretical range of h_c/h_n in this plot is $0 \leq h_c/h_n < \infty$, but it is only plotted for $0 \leq h_c/h_n < 2$ herein. Obviously, K_u at $u = h_c/h_n$ approaches zero as $h_c/h_n \to \infty$, and approaches infinity as $h_c/h_n \to 1.0$

other hand, if h_c/h_n increases from zero and approaches unity, K_u at $u = h_c/h_n$ increases correspondingly and approaches infinity as a limit. For gaining insight into the effects of h_c/h_n and N on K_u at $u = h_c/h_n$ at a glance, we use (3.89) to plot K_u at $u = h_c/h_n$ against h_c/h_n with the N-value as a parameter, as shown in Fig. 3.12. Clearly, Fig. 3.12 shows that K_u at $u = h_c/h_n$ approaches infinity as $h_c/h_n \to 1$. This trend implies that the closer the two inflection points approach unity, the larger is K_u at $u = h_c/h_n$, irrespective of the M or S profile. Figure 3.12 also shows that the smaller the N-value, the larger is K_u at $u = h_c/h_n$, tending to behave opposite to the trend of K_u everywhere along a profile except at $u = h_c/h_n$. In any case, attention should be paid to the utmost limit at $h_c/h_n = 1$, where K_u at $u = h_c/h_n$ is infinite and yet undefined because both M and S profiles collapse onto the C profile by creating two singularities at $u = h_c/h_n = 1$.

3.9 Summary

The one-dimensional approach has been traditionally used by hydraulicians to solve the GVF problem for free-surface water flow in channels. Many investigators have attempted to solve the GVF equation for h, with M, N, and h_c/h_n as the three parameters. Despite their efforts to overcome the drawback of the conventional method in the evaluation of the two integrals appearing in (3.12), our prospect to develop a viable technique to counter it without recourse to the VFF table has not been bright for decades until recently there has been the extensive and growing use of the mathematics software, such as the **Mathematica**, **Maple** and **Matlab**, to perform effectively and efficiently the integration of the improper rational function (i.e., the reciprocal of the slope of the GVF profile) via the partial-fraction expansion or the infinite-series expansion. The availability of such powerful software has helped integrate successfully the two integrals using the GHF without recourse to the VFF table. This merits attention because the GHF embraced in the GHF-based solution can henceforth replace the VFF. The GHF-based solutions can henceforth play the role of the VFF table in the interpolation of the VFF-values. So, after decades' long efforts made by hydraulicians to solve the GVF equation using exclusively the VFF to compute GVF profiles in channels with cross-sectional shape other than wide rectangle, we have finally come up with a novel approach, in which the VFF is no longer needed in the integration of the two integrals. The impact of the GHF-based solutions on the GVF computation is believed to be among the most significant results that hydraulicians have so far attained. This constitutes our principal conclusion, perhaps a milestone, if perceived from the history of the GVF computation in open channels.

In addition to our success to solve the GVF equation using the GHF, we have validated the GHF-based solution using the alternative analytical solution based on the ETF. Because both GHF-based solution and ETF-based solution are exact, the validation of one solution by the other in effect establishes the identity of both. In practice, the GHF-based solution proves to be more useful and advantageous due mainly to the orderly-behaving function parameters of the GHF (i.e., the combined expressions of M and N). The GHF-based solution is further shown to have an additional advantage attributed to the mathematical capability of the M- and N-values in the GHF to accept real numbers, embracing rational numbers, which are at the most acceptable to the ETF-based solution. The acceptance of real numbers by the GHF can relax restrictions placed on the assumed constant M- and N-values in the direct integration of the two integrals. The eventual relaxation of the assumed constant M- and N-values will thus result in extending the application of the GHF-based solution with the fixed values of M and N to the numerical computation of GVF profiles in channels with cross-sectional shape other than wide rectangle, whose M- and N-values may vary considerably, if not rapidly, with h. In our diverse perspectives, the potential to make the best use of the GHF along with its capability to vary the values of M and N through their relations with h in the computation of GVF profiles reflects a vital future phase of the principal conclusion.

Included among the other conclusions drawn from this study are some findings and minor results, which are recapped in order of presentation as follows:

1. As far as flows in wide channels are concerned, we can henceforth use (3.1) (or (3.5)) as GVF equation in search of its solution for both hydraulically smooth and fully rough flows because the solution of (3.1) for $m = 0$ (i.e., $N = 3$) for fully rough flows can be regarded as that for any m-value for hydraulically smooth flows. This may explain why the Chézy formula has been extensively used in the GVF computation in wide channels, regardless of whether flows under study are hydraulically smooth or fully rough, for $N = 3$ is applicable to both hydraulically smooth flows and fully rough flows if the Chézy formula is used in the formulation of (3.1).
2. The complete GHF-based solutions of the h_n-based dimensionless one-dimensional GVF equation may constitute any combination of (3.30) (or (3.29) and (3.31)) for $0 \leq u < 1$ with (3.43) (or (3.42) and (3.44)) for $u > 1$. By implication, any of the nine combinations can represent the complete GHF-based solutions, and all the differences in the GHF expressions among them can be reconciled by one another through conversion using the recurrence formula, (3.34), as established in Appendix B. In fact, it is merely a matter of our preference in this chapter to use a combination of (3.30) and (3.43) as typical of the complete solutions of (3.5) for $0 \leq u < \infty$, except $u = 1$.
3. Although we have obtained the GHF-based solution in the domain of $|u| > 1$ from (3.5) (i.e., a transformed form of (3.5) with its variable, w, being expressed as u^{-1}) rather than from (3.5), we have proved that (3.43) is also obtainable from (3.30) using the relation, (C.1) or (C.3), to connect one GHF in the domain of $|u| < 1$ to two GHF in the domain of $|u| > 1$. As shown in Appendix C, (C.7) transformed from (3.30) by means of (C.5) and (C.6) is identical to (3.43). In the reverse transformation, we have likewise proved that (C.10) so obtained from (3.43) through the two relations, (C.8) and (C.9), to connect one GHF in the domain of $|u| > 1$ to two GHF in the domain of $|u| < 1$, is identical to (3.30).
4. It has been shown that for the given boundary conditions we can plot the h_n-based dimensionless GVF profiles in a wide channel ($M = 3$) on all but horizontal and adverse slopes using the GHF-based solution, such as a combination of (3.30) and (3.43), with h_c/h_n and N as two parameters. For illustration, we have plotted Figure 3.1, in which the h_n-based dimensionless GVF profiles in sustaining channels for $M = 3$ and $N = 10/3$ (or $m = 1/6$) are shown against the various values of h_c/h_n.
5. All the eight h_n-based GVF profiles can be classified by demarcating u in the entire domain of u, i.e., $0 \leq u < \infty$, excluding at $u = 1$, and then described by using either (3.30) or (3.43), as tabulated in Table 3.1, except for flow on the C slope ($h_c/h_n = 1$). Unlike the M and S profiles which diverge at $u = 1$, the C1 and C3 profiles, as represented respectively by (3.50) and (3.49), neither diverge nor converge at $u = 1$ because both equations have a singularity at $u = 1$, except for $M = N = 3$. We have also proved that both C1 and C3 profiles in the

simplest case of $M = N = 3$ become continuous straight lines without a hole being created at $u = 1$.

6. We have found that the C profiles cannot be classified into three types, C1, C2, and C3, corresponding to the three zones in a channel, such as $u > h_c/h_n = 1$ (or $h > h_c = h_n$), $u = 1$ (or $h = h_c = h_n$), and $0 \leq u < h_c/h_n = 1$ (or $0 \leq h < h_c = h_n$), respectively, because the so-called C2 profile classified in the middle zone ($u = 1$) by Chow (1959) is nothing but a singularity (which is undefined), though a case for $N = 3$ may be excluded. In any event, to regard the uniform flow just only for $N = 3$ as a type of the C2 profile is unwarranted, thus excluding the C2 profile from our classification.
7. The point of inflection on a flow profile has been shown to take place at a point where du/dx_* is an extremum (i.e., either a minimum or a maximum). We have found from (3.80) that the only possible condition under which $d^2u/dx_*^2 = 0$ is (3.81) for any values of M and N. However, (3.81) becomes invalid for $h_c/h_n = 0$ because it only yields a trivial solution, i.e., $u = 0$. In particular, (3.81) for $M = 3$ and $N = 10/3$ reduces to (3.82), which is exactly the equation obtained by Gunder (1943) and Venutelli (2004).
8. We can locate two points of inflection on each M profile graphically; one on an M1 profile ($u > 1$) and the other on an M3 profile ($0 \leq u \leq h_c/h_n$). Both inflection points so determined on the M1 and M3 profiles for each of the four N-values studied, as tabulated in Table 3.7, are all expressed in terms of u, irrespective of the boundary conditions. We have also found that the closer to unity the value of h_c/h_n and/or the higher the N-value, the closer the two inflection points move from above and below the given h_c/h_n-value toward unity as their asymptotic limit. Conversely, the closer to zero the value of h_c/h_n and/or the smaller the N-value, the farther the two inflection points stay apart and eventually go to extremes, one on the M3 profiles approaching zero and the other on the M1 profiles going to infinity, as $h_c/h_n \to 0$ (see Fig. 3.11), or both disappearing, i.e., $u = 0$, if $h_c/h_n = 0$ or $N = 3$.
9. All the plotted profiles show that the higher the N-value (implying the rougher the flow and the channel boundary), the higher is the curvature of a flow profile, as illustrated in Figs. 3.2, 3.3 and 3.4, except at $u = h_c/h_n$, where the effect of the N-value on the curvature is reversed, as shown in Fig. 3.12.
10. Once we get the normal-depth-based dimensionless GVF solutions (x_*, u), we can easily obtained their corresponding dimensional GVF solutions (x, h), through the transformation relations $x = x_* h_n/S_{0*}$ $(= x_* h_n/\tan\theta)$ and $h = uh_n$.

This chapter is focused on the normalization of x and h based on h_n in the formulation of the dimensionless one-dimensional GVF equation and its solution based on GHF. Unfortunately, h_n becomes infinite for flow in horizontal channels and undefined for flow in adverse channels. Therefore, if we plan to have the dimensionless GVF equation so formulated applied to all types of channel slopes, including horizontal and adverse slopes, we should normalize x and h based on h_c rather than on h_n. However, in case that x and h are normalized based on h_c, the GHF used in the solution of the h_c-based dimensionless one-dimensional GVF equation has been found to be in a more complicated form than that shown in this chapter, thus necessitating the formulation of a different identity relation between the two contiguous GHF based on h_c. Use of such an h_c-based recurrence formula has enabled the mutual conversion of the two contiguous GHF based on h_c. This and other related issues, which we have addressed and resolved, are reported in the next chapter.

References

Abramowitz M, Stegun IA (1972) Handbook of mathematical functions with formulas, graphs, and mathemtical tabels, 10th edn. Department of Commerce, National Buereau of Standards, Applied Mathematics Series No. 55, USA.

Bresse J (1860) Cours de mécanique appliqué. hydraulique, 2e partie edn. Mallet-Bachelier, Paris

Chen CL, Wang CT (1969) Nondimensional gradually varied flow profiles. J Hydraulics Div, ASCE 95(HY5):1671–1686

Chen CL (1991) Unified theory on power laws for flow resistance. J Hydraulic Eng, ASCE 117(3):371–389

Chow VT (1957) Closure of discussions on integrating the equation of gradually varied flow. J Hydraulics Div, ASCE 83(1):9–22

Chow VT (1959) Open-channel hydraulics. McGraw-Hill, New York

Gunder DF (1943) Profile curves for open-channel flow. Transactions, ASCE 108:481–488

Jan CD, Chen CL (2012) Use of the Gaussian hypergeometric function to solve the equation of gradually-varied flow. J Hydrol 456–457:139–145

Jan CD, Chen CL (2013) Gradually varied open-channel flow profiles normalized by critical depth and analytically solved by using Gaussian hypergeometric functions. Hydrol Earth Syst Sci 17:973–987

Korn GA, Korn TM (1961) Mathematical handbook for scientists and engineers. McGraw-Hill, New York

Luke YL (1975) Mathematical functions and their approximations. Academic Press Inc, New York

Olde Daalhuis AB (2010) Hypergeometric function. In: Olver WJ et al (eds) NIST Handbook of mathematical functions. Cambridge University Press, Cambridge

Pearson CE (1974) Handbook of applied mathematics. Selected results and methods, Van Nostrand Reinhold, New York

Seaborn JB (1991) Hypergeometric functions and their applications. Springer, New York

Subramanya K (2009) Flow in open channels, 3rd edn. McGraw-Hill, Singapore

Venutelli M (2004) Direct integration of the equation of gradually varied flow. J Irrig Drainage Eng, ASCE 130(1):88–91

Whittaker ET, Watson GT (1992) An introduction to the general theory of infinite processes and of analytic functions with an account of the principal transcendental functions. Cambridge University Press, New York, A course of modern analysis

Wolfram S (1996) The mathematica book, 3rd edn. Wolfram Media and Cambridge University Press, Champaign

Zaghloul NA (1998) Hydraulic exponents m and n for gravity flow pipes. Adv Water Resour 21(3):185–191

Chapter 4
Critical-Depth-Based Dimensionless GVF Solutions Using the GHF

4.1 Introduction

As mentioned in the previous chapters, many hydraulic engineering works involve the computation of surface profiles of gradually-varied flow (GVF) that is a steady non-uniform flow in an open channel with gradually changes in its water surface elevation. The most widely used methods for computing GVF profiles could be classified into the step method and the direct integration method. The step method is a numerical method and primarily used in natural channels with non-prismatic sections. On the other hand, the direct integration method involves the integration of the GVF equation and may be performed by using analytical, semi-analytical, or numerical procedures. In some prismatic channels, such as artificial channels, the GVF equation can be simplified so as to let the analytical (or semi-analytical) direct integration can be applied. The analytical direct-integration method is straightforward and can provide the total length of the profile in a single computation step. In the direct-integration method, the one-dimensional GVF equation is usually normalized to be a simpler expression in advance so as to allow the performance of direct integration. In most cases, the GVF equation is normalized by the normal depth h_n, as shown in text books, such as that by Chow (1959), French (1986), Subramanya (2009), among others. While in some cases, it is normalized by the critical depth h_c, as done by Chen and Wang (1969), Jan and Chen (2013). Abdulrahman (2010) discussed the limit slope in uniform flow computation and developed a method to find a solution for normal depths in a rectangular channel with a given Froude number, bed slope, and roughness. The varied-flow function (VFF) that is needed in the direct-integration method conventionally used by Bakhmeteff (1932), Chow (1955, 1957, 1959) and Kumar (1978), among others, has a drawback caused by the imprecise interpolation of the VFF-values. To overcome the drawback, Jan and Chen (2012) already successfully used the Gaussian hypergeometric functions (GHF) to analytically solve the GVF equation in sustaining wide channels without recourse to the VFF. The detailed derivation and discussion of the normal-depth-based dimensionless GVF profiles solved by the GHF has been presented in Chap. 3. However, success

C.-D. Jan, *Gradually-varied Flow Profiles in Open Channels*,
Advances in Geophysical and Environmental Mechanics and Mathematics,
DOI: 10.1007/978-3-642-35242-3_4, © Springer-Verlag Berlin Heidelberg 2014

to formulate the normal-depth (h_n)-based non-dimensional GVF profiles expressed in terms of GHF for flow in sustaining channels does not necessarily warrant that it can likewise prevail to use h_n in the normalization of the GVF equation for flow in adverse channels because h_n for an assumed uniform flow in adverse channels is undefined. This chapter focuses on the direct integration method that used to analytically compute the GVF profiles in sustaining and non-sustaining channels, in which the GVF equation is normalized by using the critical depth h_c and then analytically solved by using the GHF.

4.1.1 Background and Motivation of This Study

Because the normal depth h_n for an assumed uniform flow in adverse channels is undefined, success to formulate the h_n-based dimensionless GVF profiles expressed in terms of the Gaussian hypergeometric functions (GHF) for flow in sustaining channels, as reported in Chap. 3, does not necessarily warrant that it can likewise prevail to use h_n in the normalization of the GVF equation for flow in adverse channels. Even though one can use an imaginary flow resistance coefficient to evaluate h_n for such an assumed uniform flow in adverse channels as shown in the book of Chow (1959), and as will be treated later in this chapter, it is inappropriate to use such an h_n in the normalization of the GVF equation for flow in adverse channels because h_n so evaluated is fictitious and the two normalized variables in such h_n-based dimensionless GVF equation collapse (in virtue of $h_n \to \infty$) as the channel-bottom angle of inclination, θ, approaches zero.

To fill such a mathematical gap as a result of the indiscrete choice of a characteristic length in the normalization of the GVF equation for GVF in adverse channels, Chen and Wang (1969) used the critical depth h_c to replace h_n along with the help of the traditional assumption of negligibly small bottom angle θ to derive two h_c-based dimensionless GVF equations, one for GVF in sustaining channels and the other for GVF in adverse channels. For $N = 3$ (equivalent to using Chézy formula), they obtained the exact solutions, i.e., often called the Bresse solution (Bresse 1860; Chow 1959), from both equations, but for $N = 10/3$ (equivalent to using Manning's formula) they integrated both equations numerically over the normalized depth, h/h_c, to get the h_c-based dimensionless GVF profiles in sustaining and adverse channels. Subsequently, Chen and Wang exemplified the solutions so obtained by plotting the h_c-based dimensionless GVF profiles in a series of interconnected sustaining and adverse channels, having a horizontal channel sandwiched in between them. Noteworthy in plotting such continued flow profiles on various slopes are some required "internal" boundary conditions yet to be imposed at places where the state of flow suddenly changes, such as at overfalls and hydraulic jumps.

In a case being more general than that treated above, i.e., without imposing the assumption of negligibly small θ in the formulation of the two h_c-based dimensionless GVF equations, one for GVF in sustaining channels and the other for GVF in adverse channels, one can solve both equations analytically for a given N-value for such two h_c-based dimensionless GVF profiles (which should be able to be expressed using the GHF). The method to obtain such GHF-based solutions is believed to be the same as that used in Chap. 3, from which we obtain the analytical solutions of the h_n-based dimensionless GVF profiles expressed in terms of the GHF for GVF in sustaining channels. Besides, because no assumption of negligibly small θ is required in the formulation of such two h_c-based dimensionless GVF equations, we can make the best use of the GHF-based dimensionless solutions of both equations to prove that the h_c-based dimensionless M and A profiles reduce asymptotically to the h_c-based dimensionless H profiles as $\theta \rightarrow 0$. In perspective, this study lays the theoretical foundation to compute such h_c-based dimensionless GVF profiles in a series of interconnected sustaining and adverse channels subject to a variety of internal boundary conditions.

4.1.2 Literature Survey of GVF Profiles in Adverse Channels

Before proceeding to formulate the two h_c-based dimensionless GVF equations, one for GVF in sustaining channels and the other for GVF in adverse channels, it is worthwhile to review all what have been attained by previous investigators to compute the GVF profiles in both sustaining and adverse channels. In Chap. 2, we have reviewed quite comprehensively most of the previous work on the computation of the GVF profiles in sustaining channels. Therefore, in this chapter, we can skip reappraising whatever is covered in Chap. 2, and focus on the literature survey only of the computed GVF profiles in adverse channels.

As a matter of fact, there are not many previous investigators who have contributed to the computation of the GVF profiles in adverse channels. One of the earliest investigations in this study may be credited to Matzke (1937), who formulated the h_n-based dimensionless GVF equation for flow in adverse channels, thereby integrating the indefinite integrals of a proper fraction, which are expressed from the reciprocal of the rational function representing the slope of the GVF profile. Because his GVF equation is equivalent to the one for GVF in wide channels with the flow resistance expressed by the Chézy formula, namely $M = N = 3$, the second indefinite integral merges into the first indefinite integral to form a sole indefinite integral, which he called the varied-flow function (VFF), following in Bakhmeteff's (1932) footsteps. However, unlike the four methods used by Bakhmeteff to compute the VFF-values numerically, Matzke evaluated them by a graphical method and constructed a table thereof for any N-values between 3 and 4, inclusive, for computing the h_n-based dimensionless GVF profiles in adverse channels. In addition, Matzke obtained two analytical solutions from his GVF equation for $N = 3$ and 4, using the elementary transcendental functions (ETF), which he used to cross-examine his graphical solutions of the VFF-values with their numerical errors not more than 0.33 %.

Chow (1957) was probably the first person who directly computed the VFF-values for GVF in adverse channels from the corresponding two transformed VFF-values for GVF in sustaining channels. He further tabulated the VFF-values so computed for N-values ranging from 3 to 5.5 for use in engineering applications to compute the h_n-based dimensionless GVF profiles in adverse channels (Chow 1959).

In the direct integration of the GVF equation, regardless of whether it is carried out for sustaining or adverse channels, most previous investigators expressed the rational function representing the reciprocal of the slope of the GVF profile in terms of the dimensionless flow depth rather than in terms of the geometric elements of a given channel section before the two indefinite integrals were evaluated. Nevertheless, Allen and Enever (1968) as well as Kumar (1979) did it in reverse by substituting their geometric elements of a rectangular, trapezoidal, or triangular channel section into the section factor, Z, and the conveyance, K, of the channel section in the GVF equation before evaluating the two indefinite integrals. Subsequently, they evaluated the two indefinite integrals using the partial-fraction expansion and then got the ETF-based solution for each of the cross-sectional shapes under study. Allen and Enever integrated the GVF equation for gradually varied flow in adverse channels with rectangular, triangular, especially wide or deep rectangular cross-sectional shape expressed in terms of the depth and the hypothetical normal depth, among the other relevant geometric elements of the given channel section, while Kumar did likewise for GVF in adverse channels with wide rectangular, regular rectangular, deep-narrow rectangular, and wide parabolic cross-sectional shapes. For practical applications, Kumar further constructed a table of his exclusively defined VFF for wide rectangular and wide parabolic channels as well as for regular rectangular channels with various aspect ratios.

Another approach taken by Zaghloul (1990, 1992) to integrate the GVF equation for the profiles of gradually varied flow in circular pipes was the same as that used by Allen and Enever (1968) by substituting the geometric elements of a circular conduit section into Z and K in the GVF equation before integrating it by use of Simpson's rule (Zaghloul 1990) or both direct step and integration methods (Zaghloul 1992). After all, he developed a computer model on the basis of such methods to compute GVF profiles not only in sustaining and horizontal pipes, but also in adverse pipes.

4.1.3 Objectives and Scope of This Study

For computing GVF profiles in sustaining channels, one can formulate and solve the h_n-based dimensionless GVF equation, as presented by Eq. (2.2) or (3.5). Nevertheless, for computing GVF profiles in adverse channels, one cannot do likewise due to the difficulty in quantifying the h_n-value to be used in the normalization of the GVF equation because h_n for an assumed uniform flow in adverse channels is undefined. Use of h_c instead of h_n in the normalization of the GVF equation for GVF in adverse channels can resolve such a vital issue resulting from the undefined h_n because h_c is uniquely determined for a given discharge, Q, as will be expressed later. The primary objective of this chapter is thus to formulate and then solve the two h_c-based dimensionless GVF equations, one for GVF in sustaining channels and the other for GVF in adverse channels, along with the intention to remove the conventional assumption of negligibly small θ in the formulation of both equations. The removal of such a negligibly small θ assumption made in the formulation of both equations may lead to the establishment of both equations to reduce to the h_c-based dimensionless GVF equation for GVF in horizontal channels as $\theta \rightarrow 0$. The method to obtain the GHF-based solutions from the preceding two h_c-based dimensionless GVF equations is believed to be the very same as that used in Chap. 3.

Because the Mathematica software (Wolfram 1996) can accurately output the GHF-based solutions, we will use it wherever possible in this study so as to facilitate the acquisition of the GHF-based solutions for various N-values to be studied. Each N-value can represent the mth power of the power-law flow resistance formula (Chen 1991), which will be incorporated in the formulation of the GVF equation. Upon normalization of the GVF equation based on h_c, we will solve it using the GHF in this chapter and then proceed to prove and/or analyze the following issues (that will be presented in the next chapter):

1. the asymptotic reduction of the GHF-based solution of the h_c-based dimensionless GVF equation for GVF in sustaining channels to its equivalent equation based on h_c for GVF in adverse channels, or vice versa, as $\theta \rightarrow 0$;
2. the plotting and classification of the GHF-based solutions using h_c/h_n or h_n/h_c as the parameter;
3. conversion from the h_c-based solutions to the h_n-based solutions, or vice versa;
4. the two singularities of the h_c-based slopes of the critical (C) profiles at the dimensionless critical depth (i.e. unity);
5. the inflection points of the mild (M), horizontal (H), and adverse (A) profiles; and
6. the curvature of all types of the GVF profiles.

All other related issues with regard to the normalization of the GVF equation based on h_n, such as how to acquire the GHF-based solutions thereof and the subsequent proof and/or analysis of the features and properties of the h_n-based solutions, are addressed and resolved in Chap. 3.

4.2 Formulation of the h_c-Based Dimensionless GVF Equation

Before deriving the h_c-based dimensionless GVF profiles in sustaining channels ($\theta > 0$), horizontal channels ($\theta = 0$), and adverse channels ($\theta < 0$), we first formulate subsequently the h_c-based dimensionless GVF equations for such three types of channels.

4.2.1 Dimensionless GVF Equation for Flow in Sustaining Channels

The dimensional form of the GVF differential equation expressed in terms of the flow depth prior to normalization has been shown in Eq. (1.41) or (3.1). For analysis in this chapter as a starting point, we simply write it down below without elaborating its derivation,

$$\frac{dh}{dx} = S_{0*}\frac{1-(h_n/h)^N}{1-(h_c/h)^M}. \tag{4.1}$$

This equation is a first order differential equation that denotes the relation between the flow depth h and the axial distance x along the open channel having a slope S_0. Note that no assumption of negligibly small θ is imposed in the formulation of (4.1). In Eq. (4.1), $S_{0*} = S_0/\cos\theta = \tan\theta$; h is the depth of flow measured perpendicularly from the channel bottom; x is the longitudinal coordinate measured along the channel bottom; h_n is the normal depth; h_c is the critical depth; and M and N are the hydraulic exponents for critical-flow computation and uniform-flow computation, respectively.

Though the validity of (4.1) is not at issue, one should be aware of its implication before normalizing it for solution. Firstly, the resistance to GVF in open channels, as incorporated in (4.1), is conventionally evaluated on the assumption that the resistance at any section is equal to what it would be if the same Q passed through the same section under conditions of uniformity. However, this conventional assumption made in (4.1) does not take the effect of boundary nonuniformity into account in the evaluation of the nonuniform flow resistance. According to Rouse (1965), the effect of boundary nonuniformity on the flow resistance should include any change in shape or size of the section along the longitudinal direction. Secondly, (4.1) cannot be used to compute GVF profiles in nonsustaining (horizontal or adverse) channels because h_n is infinite for flow in horizontal channels and undefined for flow in adverse channels. Thirdly, N is related to the power in the power-law flow resistance formula (Chen 1991), which has been incorporated in formulating (4.1). As given in Chap. 2, we use $N = 3$ for hydraulically smooth flows and $N = 2m + 3$ (where m is the exponent of a unified similarity variable used in the power laws of the wall) for fully rough flows. This means that if $m = 0$ for fully rough flows, $N = 2m + 3$ upon substitution of $m = 0$ reduces to $N = 3$, which is exactly the N-value derived from the Chézy formula for hydraulically smooth flows. By implication, the solution of (4.1) for $m = 0$ (i.e., $N = 3$) for fully rough flows can be treated as that for any m-value for hydraulically smooth flows. Therefore, such implication of (4.1) should be kept in mind when (4.1) is normalized for computing GVF profiles in sustaining channels.

By and large, there are two ways in which x and h in (4.1) can be normalized: One is based on the normal depth, h_n, as usually adopted by many researchers (Chow 1959) for GVF in sustaining channels, and the other based on the critical depth, h_c, as treated by Chen and Wang (1969), who adopted h_c in place of h_n to compute in one sweep GVF profiles in a series of interconnected sustaining and nonsustaining channels. Therefore, for the first case in which (4.1) is normalized based on h_n, as done in Chap. 2, we can introduce the dimensionless variables $u(= h/h_n)$ and $x_*(= xS_{0*}/h_n)$, thereby expressing (4.1) in terms of u and as

$$\frac{du}{dx_*} = \frac{1 - u^N}{u^{N-M}\left(\lambda^M - u^M\right)}, \tag{4.2}$$

which has been referred to as the dimensionless GVF equation based on h_n with λ $(= h_c/h_n)$, M, and N as three parameters. The characteristic length ratio λ in (4.2) is the primary parameter that classifies GVF profiles into those in the mild (M), critical (C), and steep (S) channels if λ is less than, equal to, and larger than unity, respectively. In fact, the direct integration of (4.2) using the Gaussian hypergeometric functions (GHF) and their analysis are the subject of Chap. 3. However, to avoid read dressing the same issues herein, we will not henceforth obtain and analyze the h_n-based GHF solutions of (4.2) unless compared otherwise with their counterparts based on h_c. In addition, the symbol u has been conventionally used to represent the h_n-based dimensionless flow depth (i.e., $u = h/h_n$) in the analysis of GVF profiles, as mentioned in Chap. 2. We will use the symbol v to represent the h_c-based dimensionless flow depth in Chaps. 4 and 5, even though it is commonly used to represent the velocity.

Instead of h_n, if we want to use h_c to normalize x and h in (4.1), we cannot simply adopt the dimensionless variables v $(= h/h_c)$ and $x_\sharp$ $(= xS_{0*}/h_c)$ to rewrite (4.1) in terms of v and $x_\sharp$ as

$$\frac{dv}{dx_\sharp} = \frac{1 - \lambda^{-N}v^{-N}}{1 - v^{-M}} = \frac{\lambda^N v^N - 1}{\lambda^N v^{N-M}(v^M - 1)}, \tag{4.3}$$

because the reciprocal of (4.3) so normalized with λ, M, and N as three parameters cannot be integrated to obtain a solution for GVF profiles in horizontal channels because the singularity in (4.3), i.e., $h_n \to \infty$, causes the parameter value of h_n/h_c $(= \lambda^{-1})$ to go to infinity (i.e., $\lambda \to 0$). To circumvent the singularity, it is necessary that the scaling factor adopted in $x_\sharp$, i.e., S_{0*}/h_c, needs to be reconciled. After exploring a few viable scaling factors by trial and error, we find that multiplying both sides of the equal sign in (4.3) by λ^N can have $x_\sharp$ redefined as xS_{c*}/h_c (in which $S_{c*} = S_c/\cos\theta$) without changing the definition of v $(= h/h_c)$, thus resulting in the following dimensionless GVF equation based on h_c with h_c/h_n, M, and N as three parameters:

$$\frac{dv}{dx_\sharp} = \frac{1 - \lambda^N v^N}{v^{N-M}(1 - v^M)}, \tag{4.4}$$

which is valid in the domain $0 \le v < \infty$.

In (4.4), the critical slope, S_c, used in redefining $x_\sharp$ is given by the discharge formula $Q = K_c S_c^q$ for a given discharge Q and the predetermined h_c (via

$Q = \sqrt{g\cos\theta/\alpha}\sqrt{A_c^3/T_c}$, where g is the gravitational acceleration, α is the energy coefficient, A_c is the cross-sectional area of critical flow, and T_c is the top width of the critical-flow section at the free surface), which is in turn used to quantify K_c, the conveyance of the channel section for critical flow. In the above discharge formula, q, the exponent of S_c is not necessarily equal to $1/2$ (see Chow 1959; Chen 1991) although $1/2$ is always adopted in the power-law flow resistance formulas for fully rough flows, such as Chézy's and Manning's formulas. It should be noted that the normalized longitudinal coordinate, $x_\sharp$ $(= xS_c/h_c)$, adopted by Chen and Wang (1969) in their h_c-based dimensionless GVF equation is similar to that used in (4.4) except that they made an additional assumption of negligible small θ in formulating the h_c-based dimensionless GVF equation, thus $\cos\theta \to 1$, which disappears from their expression of $x_\sharp$.

Because the discharge formula can be generally expressed as $Q = K_c S_c^q = K_n S_0^q$ (where K_n is the conveyance of the channel section for normal flow and S_0 is the slope of the channel bottom), we can establish the following relation.

$$\frac{S_0}{S_c} = \left(\frac{K_c}{K_n}\right)^{1/q} = \left(\frac{h_c}{h_n}\right)^{N/2q} = \lambda^{N/2q}, \tag{4.5}$$

which is valid for both hydraulically smooth and fully rough flows. As cxplained in Chap. 2, a critical analysis of the power-law resistance formula has resulted in $N/2q = 3$ for hydraulically smooth flows and $N/2q = 2m + 3$ for fully rough flows. Because $N = 3$ (or $m = 0$) for fully rough flows can be treated as that for any m-value for hydraulically smooth flows, we can further establish $N/2q = N$, from which we have $q = 1/2$ irrespective of whether the flow under study is hydraulically smooth or fully rough. Thus, (4.5) on substitution of $q = 1/2$ is expressed as

$$\frac{S_0}{S_c} = \left(\frac{K_c}{K_n}\right)^{2} = \left(\frac{h_c}{h_n}\right)^{N} = \lambda^{N}, \tag{4.6}$$

where $N = 3$ for hydraulically smooth flows and $N = 2m+3$ for fully rough flows. It is noted that (4.6) is also valid for GVF profiles in horizontal channels because all the expressions in (4.6) can reduce to zero by virtue of $S_o = 0$, $K_n \to \infty$, and $h_n \to \infty$.

Once (4.2) and (4.4) are established, we can, respectively, relate the normalized flow depth and the normalized longitudinal coordinate defined on the basis of h_n to their counterparts defined on the basis of h_c as follows:

$$\frac{u}{v} = \frac{h_c}{h_n} = \lambda, \tag{4.7}$$

$$\frac{x_*}{x_\sharp} = \frac{h_c}{h_n}\frac{S_{0*}}{S_{c*}} = \frac{h_c}{h_n}\frac{S_0}{S_c} = \left(\frac{h_c}{h_n}\right)^{N+1} = \lambda^{N+1}. \tag{4.8}$$

It is worthwhile to point out that the right-hand expression of the second equal sign in (4.8) has resulted from the relation appearing in (4.6). As can be proved later, both normalized variables so related in (4.7) and (4.8) will be useful in converting the solutions of (4.2) to its counterparts obtained from (4.4), or vice versa.

4.2.2 Dimensionless GVF Equation for Flow in Horizontal Channels

Although Chow (1959) did not classify horizontal channels as a type of sustaining channels, we can prove that the h_c-based dimensionless GVF equation, (4.4), is still valid for GVF profiles in horizontal channels. In other words, for GVF profiles in horizontal channels, because we have $\theta = 0$, $h_n \to \infty$, and $h_c/h_n \to 0$ (i.e., $\lambda \to 0$), (4.4) can be simplified to

$$\frac{dv}{dx_\sharp} = \frac{v^{M-N}}{1 - v^M} \tag{4.9}$$

which is still valid in the domain of $0 \leq v < \infty$. Equation (4.9) is the dimensionless equation based on h_c for GVF profiles in horizontal channels with M and N as two parameters. For convenience, (4.9) may be referred to as the asymptote of (4.4) as $h_c/h_n \to 0$. Under the assumption of constant M and N, (4.4) can be analytically integrated over v using the GHF, as will be shown later, while we can directly integrate (4.9) without resort to the GHF in the following manner. The reciprocal of (4.9) upon rearrangement is expressible as

$$\frac{dx_\sharp}{dv} = v^{N-M} - v^N, \tag{4.10}$$

which, provided M and N are constant, can readily be integrated over v to yield

$$x_\sharp = \frac{v^{N-M+1}}{N-M+1} - \frac{v^{N+1}}{N+1} + \text{Const.}. \tag{4.11}$$

Equation (4.11) describes the h_c-based dimensionless GVF profiles in wide horizontal channels ($M = 3$). The constant of integration, "Const.", in (4.11) is determined from a given boundary condition. It should be noted that (4.11) can readily be identified with equation (10–26) in the book of Chow (1959), although his would-be $x_\sharp$ can be written as xS_c/h_c, which is the same expression that Chen and Wang (1969) adopted owing to the negligibly small θ assumed in the formulation of the GVF equation.

The profile classification by Chow (1959) to distinguish a horizontal slope ($\theta = 0$) from a sustaining slope ($\theta > 0$) or an adverse slope ($\theta < 0$) seems at odds with the limit of differential calculus, because the h_c-based dimensionless horizontal profiles,

(4.11), can be derived from the h_c-based dimensionless equation for GVF profiles in sustaining or adverse channels having any real θ-values, including zero. In fact, Chow could have treated GVF profiles on a horizontal slope as the asymptote of those on a sustaining slope as $\theta \rightarrow 0$, as deduced above from (4.4) to (4.11) by letting λ ($= h_c/h_n$) $\rightarrow 0$ (or $h_n \rightarrow \infty$). We shall also later prove that the GHF-based solutions of (4.4) for GVF on the mild slope (where $0 < h_c/h_n < 1$) can reduce to (4.11) when we approach its asymptote by letting $h_c/h_n \rightarrow 0$ (or $h_n \rightarrow \infty$). Likewise, the h_c-based dimensionless GVF equation for GVF in adverse channels, as will be formulated later, can also approach its asymptote, namely (4.11), as $h_c/h_n \rightarrow 0$ (or $h_n \rightarrow \infty$). Therefore, for coherence in treating the asymptote of the h_c-based dimensionless GVF equation and its solutions throughout the chapter, we henceforth regard the horizontal slope as an intermediate (or interface) between the sustaining slope and the adverse slope as θ approaches zero from either slope.

4.2.3 Dimensionless GVF Equation for Flow in Adverse Channels

The GVF equation for flow in adverse channels is described by Eq. (1.42), using a fictitious normal depth. Based on (1.42), using the same derivation procedure and using the same dimensionless variables, we can obtain the h_c-based dimensionless GVF equation for GVF in adverse channels as

$$\frac{dv}{dx_\sharp} = \frac{1 + \lambda^N v^N}{v^{N-M}(1 - v^M)}. \tag{4.12}$$

As pointed out earlier, (4.12) can reduce asymptotically to (4.9) for flow in a horizontal channel as $\theta \rightarrow 0$ [i.e., $h_n \rightarrow \infty$ or λ ($= h_c/h_n$) $\rightarrow 0$].

Undoubtedly, the issue here is the controversial approach taken in evaluating h_n for use in (4.12) because the flow resistance coefficient in the power-law resistance formula is assumed to be imaginary and the h_n-value so evaluated is fictitious in an adverse channel. Nevertheless, for a given Q, the h_c-value so computed is unique, regardless of whether it is in a sustaining or adverse channel. The parameter, λ^N, appearing in (4.12) signifies the slope ratio, S_0/S_c, as shown in (4.6), thus implying that λ^N is of a negative value for GVF in adverse ($S_0 < 0$) channels. Therefore, $h_c/h_n[= (S_0/S_c)^{1/N}]$ as a parameter for GVF in adverse channels is not so physically meaningful as that in (4.4) for GVF in sustaining channels, in which the flow criteria of the λ-value being less than, equal to, and greater than unity can represent, respectively, the GVF on mild, critical, and steep slopes. In contrast, the fictitious λ-value in (4.12) for GVF in adverse channels is nothing but an index. Although the ultimate usefulness of (4.12) may not be realized until after this issue in the evaluation of h_n is resolved, we presently rest content with such fictitious h_n values because there appears no other approach better than that taken above to evaluate h_n for GVF in adverse channels.

Although the h_n-value so evaluated is fictitious in an adverse channel regardless of whether (4.1) is normalized on the basis of h_c or h_n, we prefer to normalize (4.1) based on h_c rather than based on h_n for the following main reason: Most previous investigators, such as Chow (1957, 1959), formulated the h_n-based dimensionless GVF equation for computing GVF profiles in adverse channels, i.e., the h_n-based counterpart of (4.12) instead of the h_c-based (4.12). However, (4.12) has obvious advantages over its h_n-based counterpart because the latter cannot reduce asymptotically to (4.9) as $\theta \to 0$ (i.e., $h_n \to \infty$ or $h_c/h_n \to 0$) unless the normalizing quantity of the h_n-based GVF equation for adverse channels can be altered from h_n to h_c so as to enable one to switch the normalized variables from u to v via (4.7) and from x_* to $x_\sharp$ via (4.8) as $\theta \to 0$. By implication, if the h_n-based counterpart of (4.12) would have been used to compute GVF profiles in adverse channels, the horizontal slope could not have been treated as an intermediate (or interface) between the sustaining slope and the adverse slope as $\theta \to 0$.

4.3 GHF-Based Solutions of the h_c-Based Dimensionless GVF Equations

There are two ways to solve the h_n-based dimensionless GVF equation analytically, as treated in Chap. 3: One is based on the Gaussian hypergeometric functions (GHF) and the other on the elementary transcendental functions (ETF). We have verified in Chap. 3 that the analytical solution obtained using the GHF is identical to that using ETF. There is no question that we can likewise verify the identity of both GHF-based and ETF-based solutions to be obtained from the h_c-based dimensionless GVF equation. However, because we prove in Chap. 3 that the GHF-based solutions obtained from the h_n-based dimensionless GVF equation are more useful and versatile than the ETF-based solutions, in which the values of M and N can at most accept rational numbers, but not real numbers, this chapter is henceforth focused only on the acquisition and analysis of the GHF-based solution from the h_c-based dimensionless GVF equation.

Analogous to the direct integration of the h_n-based dimensionless GVF equation, as treated in Chap. 3, we can integrate both (4.4) and (4.12) directly and express their solutions in terms of the GHF. Before solving them, we transpose them to their reciprocals and then proceed to express each of both reciprocals as a polynomial plus a proper fraction by the process of long division (i.e., "division algorithm"). Two options are available to arrange the two indefinite integrals of the proper fraction. In principle, the two GHF-based solutions obtained by adding the two indefinite integrals together using either option are presumed to be identical. This presumption lays the foundation for establishing a functional identity relation (or often called a recurrence formula) between the two contiguous GHF, as already induced from the definition of the GHF in Appendix B. The recurrence formula so established between the two contiguous GHF is valuable in linking two unlikely related GHF. The roles of such a recurrence formula playing in the conversion of the two GHF-based solutions, among other related issues, are reported subsequently.

4.3.1 Alternative Forms of GVF Equations for $| \lambda v | > 1$

The validity of the solutions of (4.4) or (4.12) to be expressed in terms of the GHF is confined to the convergence criterion of the GHF, i.e., $| \lambda v |< 1$. Therefore, an alternative form of (4.4) or (4.12) has yet to be formulated and its solutions expressed in terms of the reconciled GHF, which can be made valid for $| \lambda v |> 1$. Every GHF-based solution of (4.4) or (4.12) thus consists of two parts: one in the domain of $| \lambda v |< 1$ and the other in the domain of $| \lambda v | > 1$. In this study, we seek two sets of two GHF-based solutions for each of (4.4) and (4.12) in three steps. In each step, we first carry out the required manipulations for (4.4) and then repeat the same treatments for (4.12) as follows: In the first step, we will formulate the alternative form of (4.4), and proceed to arrange the two indefinite integrals in both (4.4) and its alternative form in two ways. In the second step, upon derivation of two sets of the two GHF-based solutions from (4.4) and its alternative form, we will establish two recurrence formulas between each set of the two corresponding contiguous GHF. In the third step, we will validate such recurrence formulas using the general rule established in Appendix B. All the recurrence formulas so derived can be used to prove the identity of the four parametrically different, yet mutually related GHF-based solutions of (4.4) or (4.12).

To derive an alternative form of (4.4) in order for its GHF-based solutions to be valid for $| \lambda v |> 1$ for GVF in sustaining channels, (4.4) on substitution of $v = w^{-1}$ and $dv = -w^{-2}dw$ yields

$$\frac{dw}{dx_\sharp} = \frac{-\lambda^N w^2 + w^{N+2}}{1 - w^M}. \tag{4.13}$$

Obviously, the analytical solutions of (4.13), if expressible in terms of the GHF, should be valid for$| (h_n/h_c)w |< 1$ (or the reciprocal of its convergence criterion, i.e. $| (h_c/h_n)v | > 1$, upon substitution of $w = v^{-1}$). By the same token, for GVF in adverse channels, (4.12) on substitution of $v = w^{-1}$ and $dv = -w^{-2}dw$ yields

$$\frac{dw}{dx_\sharp} = \frac{\lambda^N w^2 + w^{N+2}}{1 - w^M}. \tag{4.14}$$

4.3.2 Integration of the GVF Equation for Flow in Sustaining Channels

(1) For $| \lambda v |< 1$:

For getting the h_c-based dimensionless GVF profiles in sustaining channels ($\theta > 0$) expressed in terms of the GHF from (4.4) for $| \lambda v |< 1$ and (4.13) for $| \lambda v |> 1$, we first rearrange (4.4) into its reciprocal form as

$$\frac{dx_\sharp}{dv} = \frac{v^{N-M} - v^N}{1-(\lambda v)^N} = \frac{v^{N-M}}{1-(\lambda v)^N} - \frac{v^N}{1-(\lambda v)^N}. \tag{4.15}$$

The right-hand side of (4.15) is a rational function of v, which by the process of long division can be expressed in the form of a polynomial plus a proper fraction in two ways before integration. One way is to divide the first term of the numerator on the right-hand side of (4.15) by the denominator, namely

$$\frac{dx_\sharp}{dv} = \frac{v^{N-M} - v^N}{1-(\lambda v)^N} = v^{N-M} - \frac{v^N}{1-(\lambda v)^N} + \frac{\lambda^N v^{2N-M}}{1-(\lambda v)^N}. \tag{4.16}$$

The other way is to divide the second term of the numerator on the right-hand side of (4.15) by the denominator, i.e.,

$$\frac{dx_\sharp}{dv} = \frac{v^{N-M} - v^N}{1-(\lambda v)^N} = -v^N + \frac{v^{N-M}}{1-(\lambda v)^N} - \frac{\lambda^N v^{2N}}{1-(\lambda v)^N}. \tag{4.17}$$

Integrating (4.15), (4.16) and (4.17) with respect to dv, respectively, yields the following three possible representations of the solution of (4.4) by the direct integration method.

$$x_\sharp = \int_0^v \frac{v^{N-M}}{1-(\lambda v)^N} dv - \int_0^v \frac{v^N}{1-(\lambda v)^N} dv + \text{Const.}, \tag{4.18}$$

$$x_\sharp = \frac{v^{N-M+1}}{N-M+1} + \lambda^N \int_0^v \frac{v^{2N-M}}{1-(\lambda v)^N} dv - \int_0^v \frac{v^N}{1-(\lambda v)^N} dv + \text{Const.}, \tag{4.19}$$

$$x_\sharp = \int_0^v \frac{v^{N-M}}{1-(\lambda v)^N} dv - \frac{v^{N+1}}{N+1} - \lambda^N \int_0^v \frac{v^{2N}}{1-(\lambda v)^N} dv + \text{Const.}. \tag{4.20}$$

These three GHF-based solutions to be obtained from (4.18) to (4.20) should be identical. Therefore, equating any two of such three GHF-based solutions will thus establish two recurrence formulas between the two contiguous GHF that will be discussed later.

(2) For $|\lambda v| > 1$:

Likewise, (4.13) for $|\lambda v| > 1$ upon rearrangement into its reciprocal form yields

$$\frac{dx_\sharp}{dw} = \frac{\lambda^{-N}(-w^{-2} + w^{M-2})}{1-(w/\lambda)^N}. \tag{4.21}$$

The right-hand side of (4.21) is also a rational function of w, which by the process of long division can also be expressed in the form of a polynomial plus a proper fraction in two ways before we integrate them. One way is again to divide the first term of the numerator on the right-hand side of (4.21) by the denominator, i.e.,

$$\begin{aligned}\frac{dx_\sharp}{dw} &= \frac{\lambda^{-N}(-w^{-2}+w^{M-2})}{1-(w/\lambda)^N} \\ &= -\lambda^{-N}w^{-2} - \frac{\lambda^{-2N}w^{N-2}}{1-(w/\lambda)^N} + \frac{\lambda^{-N}w^{M-2}}{1-(w/\lambda)^N}.\end{aligned} \tag{4.22}$$

The other way is again to divide the second term of the numerator on the right-hand side of (4.21) by the denominator, i.e., to write

$$\begin{aligned}\frac{dx_\sharp}{dw} &= \frac{\lambda^{-N}(-w^{-2}+w^{M-2})}{1-(w/\lambda)^N} \\ &= \lambda^{-N}w^{M-2} - \frac{\lambda^{-N}w^{-2}}{1-(w/\lambda)^N} + \frac{\lambda^{-2N}w^{N+M-2}}{1-(w/\lambda)^N}.\end{aligned} \tag{4.23}$$

Integrating (4.21), (4.22) and (4.23) with respect to dw, respectively, yields the following three possible representations of the solution of (4.13) by the direct integration method.

$$\begin{aligned}x_\sharp &= -\lambda^{-N}\int_0^w \frac{w^{-2}}{1-(w/\lambda)^N}dw \\ &\quad +\lambda^{-N}\int_0^w \frac{w^{M-2}}{1-(w/\lambda)^N}dw + \text{Const.},\end{aligned} \tag{4.24}$$

$$\begin{aligned}x_\sharp &= \lambda^{-N}w^{-1} - \lambda^{-2N}\int_0^w \frac{w^{N-2}}{1-(w/\lambda)^N}dw \\ &\quad +\lambda^{-N}\int_0^w \frac{w^{M-2}}{1-(w/\lambda)^N}dw + \text{Const.},\end{aligned} \tag{4.25}$$

$$\begin{aligned}x_\sharp &= -\lambda^{-N}\int_0^w \frac{w^{-2}}{1-(w/\lambda)^N}dw + \frac{\lambda^{-N}w^{M-1}}{M-1} \\ &\quad +\lambda^{-2N}\int_0^w \frac{w^{N+M-2}}{1-(w/\lambda)^N}dw + \text{Const.}.\end{aligned} \tag{4.26}$$

These three GHF-based solutions to be obtained from (4.24) to (4.26) should be identical; therefore, equating any two of such three GHF-based solutions will establish another two recurrence formulas between the two contiguous GHF.

4.3.3 *GHF-Based Solutions of GVF Profiles in Sustaining Channels*

The complete GHF-based solutions of the dimensionless GVF equation, (4.4), for GVF in sustaining channels are obtained by executing the integration of the two integrals in each of (4.18), (4.19) and (4.20) for $|\ \lambda v\ | < 1$ and then carrying out the same in each of (4.24), (4.25) and (4.26) for $| w/\lambda | < 1$ (or $| \lambda v | > 1$ after substituting $w = v^{-1}$ into $| w/\lambda | < 1$). It should be noted that such GHF-based solutions exclude points where $| \lambda v | = 1$ (or $| w/\lambda | = 1$) because the GHF diverges at $| \lambda v | = 1$ (or $| w/\lambda | = 1$). Physically, such points correspond to the convergent limits of the GHF in the GHF-based solutions of (4.4) at $x_\sharp = \pm\infty$, where GVF profiles run parallel with the channel bottom.

(1) For $| \lambda v | < 1$:

For $| \lambda v | < 1$, there are only four different integrals in Eqs. (4.18), (4.19) and (4.20), even though there are two different integrals in each of these equations. A general solution for these four integrals in terms of GHF can be obtained with the help of the Mathematica software (or the Maple software),

$$\int \frac{v^\phi}{1-(\lambda v)^N} dv = \frac{v^{\phi+1}}{\phi+1} g\left(\frac{\phi+1}{N}, (\lambda v)^N\right) + \text{Const.}, \tag{4.27}$$

in which $g(b, z)$ is a simplified expression of GHF with parameter $b = (\phi + 1)/N$ and the variable $z = (\lambda v)^N$. The parameter $\phi = N, 2N - M, N - M$, and $2N$, respectively, for the above mentioned four different integrals in (4.18) to (4.20), and the integration results of these four integrals in terms of GHF are

$$\int_0^v \frac{v^N}{1-(\lambda v)^N} dv = \frac{v^{N+1}}{N+1} g\left(\frac{N+1}{N}, (\lambda v)^N\right), \tag{4.28}$$

$$\int_0^v \frac{v^{2N-M}}{1-(\lambda v)^N} dv = \frac{v^{2N-M+1}}{2N-M+1} g\left(\frac{2N-M+1}{N}, (\lambda v)^N\right), \tag{4.29}$$

$$\int_0^v \frac{v^{N-M}}{1-(\lambda v)^N} dv = \frac{v^{N-M+1}}{N-M+1} g\left(\frac{N-M+1}{N}, (\lambda v)^N\right), \tag{4.30}$$

$$\int_0^v \frac{v^{2N}}{1-(\lambda v)^N} dv = \frac{v^{2N+1}}{2N+1} g\left(\frac{2N+1}{N}, (\lambda v)^N\right). \tag{4.31}$$

Replacing the two integral terms in each of (4.18), (4.19) and (4.20) using the GHF-based solutions listed in (4.28), (4.29), (4.30) and (4.31), we can express these three possible representations of the solution of (4.4) as

$$\begin{aligned} x_\sharp = & \frac{v^{N-M+1}}{N-M+1} g\left(\frac{N-M+1}{N}, (\lambda v)^N\right) \\ & - \frac{v^{N+1}}{N+1} g\left(\frac{N+1}{N}, (\lambda v)^N\right) + \text{Const.}, \end{aligned} \tag{4.32}$$

$$x_{\sharp} = \frac{v^{N-M+1}}{N-M+1} + \frac{\lambda^N v^{2N-M+1}}{2N-M+1} g\left(\frac{2N-M+1}{N}, (\lambda v)^N\right) - \frac{v^{N+1}}{N+1} g\left(\frac{N+1}{N}, (\lambda v)^N\right) + \text{Const.}, \tag{4.33}$$

$$x_{\sharp} = \frac{v^{N-M+1}}{N-M+1} g\left(\frac{N-M+1}{N}, (\lambda v)^N\right) - \frac{v^{N+1}}{N+1} - \frac{\lambda^N v^{2N+1}}{2N+1} g\left(\frac{2N+1}{N}, (\lambda v)^N\right) + \text{Const.}. \tag{4.34}$$

The above three GHF-based solutions are all valid only for $\mid \lambda v \mid < 1$. Because these three GHF-based solutions so obtained from (4.18) to (4.20), i.e., (4.32) to (4.34), should be identical, comparing any two of them establishes the following two identities:

$$g\left(\frac{N+1}{N}, (\lambda v)^N\right) = 1 + \frac{(N+1)(\lambda v)^N}{2N+1} g\left(\frac{2N+1}{N}, (\lambda v)^N\right), \tag{4.35}$$

$$g\left(\frac{N-M+1}{N}, (\lambda v)^N\right) = 1 + \frac{(N-M+1)(\lambda v)^N}{2N-M+1} g\left(\frac{2N-M+1}{N}, (\lambda v)^N\right). \tag{4.36}$$

The above two formulas are the recurrence formulas of the two contiguous GHF of the variable, $(\lambda v)^N$, with the combined parameters of M and N shown in (4.35) and (4.36). It merits attention to mention that the above recurrence formulas are only two special cases of the general identity relation of the two contiguous GHF, as shown and proved in Appendix B.

(2) For $\mid \lambda v \mid > 1$:

For $\mid \lambda v \mid > 1$ (or $\mid w/\lambda \mid < 1$ after substituting $v = w^{-1}$ into $\mid \lambda v \mid > 1$), there are four different integrals in (4.24), (4.25) and (4.26). We can also get a general solution for these four integrals in terms of GHF with the help of the Mathematica software, viz.,

$$\int \frac{w^{\phi}}{1-(w/\lambda)^N} dw = \frac{w^{\phi+1}}{\phi+1} g\left(\frac{\phi+1}{N}, (w/\lambda)^N\right) + \text{Const.}. \tag{4.37}$$

Except the difference between the variables $(\lambda v)^N$ and $(w/\lambda)^N$, the expression of (4.37) is exactly the same as (4.27). The parameter $\phi = N-2$, $M-2$, -2, and $M+N-2$, respectively, for the four integrals in (4.24), (4.25) and (4.26). The integration results in terms of GHF are

$$\int_0^w \frac{w^{N-2}}{1-(w/\lambda)^N}dw = \frac{w^{N-1}}{N-1}g\left(\frac{N-1}{N}, (w/\lambda)^N\right), \tag{4.38}$$

$$\int_0^w \frac{w^{M-2}}{1-(w/\lambda)^N}dw = \frac{w^{M-1}}{M-1}g\left(\frac{M-1}{N}, (w/\lambda)^N\right), \tag{4.39}$$

$$\int_0^w \frac{w^{-2}}{1-(w/\lambda)^N}dw = \frac{w^{-1}}{-1}g\left(\frac{-1}{N}, (w/\lambda)^N\right), \tag{4.40}$$

$$\int_0^w \frac{w^{N+M-2}}{1-(w/\lambda)^N}dw = \frac{w^{N+M-1}}{N+M-1}g\left(\frac{N+M-1}{N}, (w/\lambda)^N\right). \tag{4.41}$$

Replacing the two integrals in each of (4.24), (4.25) and (4.26) using the GHF-based solutions as listed in (4.38), (4.39), (4.40) and (4.41) we can express three possible representations of the solution of (4.13) as

$$\begin{aligned} x_\sharp = {} & \lambda^{-N} w^{-1} g\left(-\frac{1}{N}, (w/\lambda)^N\right) \\ & + \frac{\lambda^{-N} w^{M-1}}{M-1} g\left(\frac{M-1}{N}, (w/\lambda)^N\right) + \text{Const.}, \end{aligned} \tag{4.42}$$

$$\begin{aligned} x_\sharp = {} & \lambda^{-N} w^{-1} - \frac{\lambda^{-2N} w^{N-1}}{N-1} g\left(\frac{N-1}{N}, (w/\lambda)^N\right) \\ & + \frac{\lambda^{-N} w^{M-1}}{M-1} g\left(\frac{M-1}{N}, (w/\lambda)^N\right) + \text{Const.}, \end{aligned} \tag{4.43}$$

$$\begin{aligned} x_\sharp = {} & \lambda^{-N} w^{-1} g\left(-\frac{1}{N}, (w/\lambda)^N\right) + \frac{\lambda^{-N} w^{M-1}}{M-1} \\ & + \lambda^{-2N} \frac{w^{N+M-1}}{N+M-1} g\left(\frac{N+M-1}{N}, (w/\lambda)^N\right) + \text{Const.}. \end{aligned} \tag{4.44}$$

These GHF-based solutions are valid for $\mid w/\lambda \mid < 1$. Because these three GHF-based solutions so obtained from (4.24) to (4.26), i.e., (4.42) to (4.44), should be identical, comparing any two of them establishes the following two identities:

$$g\left(-\frac{1}{N}, (w/\lambda)^N\right) = 1 - \frac{(w/\lambda)^N}{N-1} g\left(\frac{N-1}{N}, (w/\lambda)^N\right), \tag{4.45}$$

$$g\left(\frac{M-1}{N}, (w/\lambda)^N\right) = 1 + \frac{(M-1)(w/\lambda)^N}{N+M-1} g\left(\frac{N+M-1}{N}, (w/\lambda)^N\right). \tag{4.46}$$

To express (4.42), (4.43) and (4.44) in terms of v $(= w^{-1})$, we can obtain the following corresponding equations

$$\begin{aligned} x_{\sharp} &= \lambda^{-N} v g\left(-\frac{1}{N}, (\lambda v)^{-N}\right), \\ &+ \frac{\lambda^{-N} v^{-M+1}}{M-1} g\left(\frac{M-1}{N}, (\lambda v)^{-N}\right) + \text{Const.}, \end{aligned} \tag{4.47}$$

$$\begin{aligned} x_{\sharp} &= \lambda^{-N} v - \frac{\lambda^{-2N} v^{-M+1}}{N-1} g\left(\frac{N-1}{N}, (\lambda v)^{-N}\right) \\ &+ \frac{\lambda^{-N} v^{-M+1}}{M-1} g\left(\frac{M-1}{N}, (\lambda v)^{-N}\right) + \text{Const.}, \end{aligned} \tag{4.48}$$

$$\begin{aligned} x_{\sharp} &= \lambda^{-N} v g\left(-\frac{1}{N}, (\lambda v)^{-N}\right) + \frac{\lambda^{-N} v^{-M+1}}{M-1} \\ &+ \frac{\lambda^{-2N} v^{-N-M+1}}{N+M-1} g\left(\frac{N+M-1}{N}, (\lambda v)^{-N}\right) + \text{Const.}. \end{aligned} \tag{4.49}$$

These are valid for $\mid \lambda v \mid > 1$. Because these three GHF-based solutions in terms of v, i.e., (4.47), (4.48) and (4.49) should be identical, comparing any two of them establishes the following two identities.

$$g\left(-\frac{1}{N}, (\lambda v)^{-N}\right) = 1 - \frac{(\lambda v)^{-N}}{N-1} g\left(\frac{N-1}{N}, (\lambda v)^{-N}\right), \tag{4.50}$$

$$g\left(\frac{M-1}{N}, (\lambda v)^{-N}\right) = 1 + \frac{(M-1)(\lambda v)^{-N}}{N+M-1} g\left(\frac{N+M-1}{N}, (\lambda v)^{-N}\right). \tag{4.51}$$

It is noted that (4.50) and (4.51) are both identical to (4.45) and (4.46), respectively, because they can be transformed to each other via $w = v^{-1}$. Likewise, it merits attention to point out that (4.50) or (4.51) is also a special case of the general identity relation between the two contiguous GHF, as shown in Appendix B.

Although the absolute value of the variable in the GHF, i.e., $\mid \lambda v \mid$, has been imposed to derive the GHF-based solutions of (4.4) for GVF profiles in sustaining channels, the complete GHF-based solutions of (4.4) can only cover the physically possible domain of λv, i.e. $0 \leq \lambda v < \infty$, thus consisting of (4.32) [or (4.33) and (4.34)] for $0 \leq \lambda v < 1$, $x_{\sharp} = \pm\infty$ at $\lambda v = 1$, and (4.47) [or (4.48) and (4.49)] for $\lambda v > 1$. Thus, we can lift the absolute-value restriction imposed on $(\lambda v)^N$ or $(\lambda v)^{-N}$ in expressing the GHF-based solutions without loss of generality. Thus, excluding $x_{\sharp} = \pm\infty$ at $\lambda v = 1$, the complete GHF-based solutions of (4.4) should consist of any combination of (4.32) [or (4.33) and (4.34)] for $0 \leq \lambda v < 1$ with (4.47) [or (4.48) and (4.49)] for $\lambda v > 1$. By implication, any of the nine combinations can represent the complete GHF-based solutions of (4.4) for $0 \leq \lambda v < \infty$ at $\lambda v = 1$ where $x_{\sharp} = \pm\infty$. Any differences in the GHF-expressions among them can be reconciled through transformation using the recurrence formulas, (4.35), (4.36), (4.50) and/or (4.51). To facilitate a comparison between the h_c-based GHF solutions of (4.4) and their counterparts based on h_n (which we have obtained in Chap. 3), we select the combination of (4.32) and (4.47) as typical of the complete solutions of (4.4) spanning $0 \leq \lambda v < 1$ and $\lambda v > 1$, respectively, unless a circumstance in favor of the other combinations arises otherwise.

4.3.4 Alternative Method to Get the GVF Profiles in Sustaining Channels

Alternatively, we can get (4.47) through an established relation connecting one GHF in the domain of the variable $\mid z \mid < 1$ to two GHF in the domain of the variable $\mid z \mid > 1$, such as (F.1) or (F.2) in Appendix F. In other words, such a relation enables one to transform the GHF-based solution of (4.4) in the domain of $\mid \lambda v \mid < 1$, i.e., (4.32), to the GHF-based solution of (4.4) in the domain of $\mid \lambda v \mid > 1$, i.e., (4.47). Although we have thus far obtained the GHF-based solution of (4.4) in the domain of $\mid \lambda v \mid > 1$ from (4.13) [i.e., a transformed form of (4.4) with its variable, w, being expressed as v^{-1}] rather than through (4.4), we can prove that (4.47) is also obtainable from (4.32) using the two relations, i.e. (F.3) and (F.4), to connect one GHF in the domain of $\mid \lambda v \mid < 1$ to two GHF in the domain of $\mid \lambda v \mid > 1$. It is proved in Appendix F that (F.5) so transformed from (4.32) through (F.3) and (F.4) is identical to (4.47).

4.4 GHF-Based Dimensionless GVF Profiles in Adverse Channels

4.4.1 Integrations of the GVF Equations for Flow in Adverse Channels

(1) For $\mid \lambda v \mid < 1$:

The h_c-based dimensionless GVF equation for flow in adverse channels has been shown in (4.12). For deriving the GHF-based solutions of (4.12) for $\mid \lambda v \mid < 1$, we rearrange (4.12) into its reciprocal form

$$\frac{dx_{\sharp}}{dv} = \frac{v^{N-M} - v^N}{1 + (\lambda v)^N}. \tag{4.52}$$

Analogous to the decomposition of (4.15) for GVF in sustaining channels into two terms by the process of long division before integrating them, there are two ways to express the right-hand side of (4.52) in the form of a polynomial plus a proper fraction. One way is to divide the first term of the numerator on the right-hand side of (4.52) by the denominator, i.e.,

$$\frac{dx_\sharp}{dv} = \frac{v^{N-M} - v^N}{1+(\lambda v)^N} = v^{N-M} - \frac{v^N}{1+(\lambda v)^N} - \frac{\lambda^N v^{2N-M}}{1+(\lambda v)^N}. \tag{4.53}$$

The other way is to divide the second term of the numerator on the right-hand side of (4.52) by the denominator, i.e.,

$$\frac{dx_\sharp}{dv} = \frac{v^{N-M} - v^N}{1+(\lambda v)^N} = -v^N + \frac{v^{N-M}}{1+(\lambda v)^N} + \frac{\lambda^N v^{2N}}{1+(\lambda v)^N}. \tag{4.54}$$

Integrating (4.52), (4.53) and (4.54) with respect to dv, respectively, yields

$$x_\sharp = \int_0^v \frac{v^{N-M}}{1+(\lambda v)^N} dv - \int_0^v \frac{v^N}{1+(\lambda v)^N} dv + \text{Const.}, \tag{4.55}$$

$$x_\sharp = \frac{v^{N-M+1}}{N-M+1} - \lambda^N \int_0^v \frac{v^{2N-M}}{1+(\lambda v)^N} dv - \int_0^v \frac{v^N}{1+(\lambda v)^N} dv + \text{Const.}, \tag{4.56}$$

$$x_\sharp = \int_0^v \frac{v^{N-M}}{1+(\lambda v)^N} dv - \frac{v^{N+1}}{N+1} + \lambda^N \int_0^v \frac{v^{2N}}{1+(\lambda v)^N} dv + \text{Const.}. \tag{4.57}$$

As mentioned earlier, the three GHF-based solutions to be obtained from (4.55) to (4.57) should be identical. Upon integration and then being expressed in terms of the GHF, comparison of these three GHF-based solutions can be equated to establish another two recurrence formulas between the two contiguous GHF.

(2) For $\mid \lambda v \mid > 1$:

Likewise, we can derive the GHF-based solutions of (4.14) for $\mid \lambda v \mid > 1$ by rearranging (4.14) into its reciprocal form as

$$\frac{dx_\sharp}{dw} = \frac{\lambda^{-N}(w^{-2} - w^{M-2})}{1+(w/\lambda)^N}. \tag{4.58}$$

The right-hand side of (4.58) is also a rational function of w, which again by the process of long division can be expressed in the form of a polynomial plus a proper fraction in two ways before integrating them. One way is to divide the first term of the numerator on the right-hand side of (4.58) by the denominator, i.e.,

$$\begin{aligned} \frac{dx_\sharp}{dw} &= \frac{\lambda^{-N}(w^{-2} - w^{M-2})}{1+(w/\lambda)^N} \\ &= \lambda^{-N} w^{-2} - \frac{\lambda^{-2N} w^{N-2}}{1+(w/\lambda)^N} - \frac{\lambda^{-N} w^{M-2}}{1+(w/\lambda)^N}. \end{aligned} \tag{4.59}$$

The other way is to divide the second term of the numerator on the right-hand side of (4.58) by the denominator, i.e.,

$$\begin{aligned}\frac{dx_\sharp}{dw} &= \frac{\lambda^{-N}(w^{-2} - w^{M-2})}{1 + (w/\lambda)^N} \\ &= \lambda^{-N}w^{M-2} + \frac{\lambda^{-N}w^{-2}}{1 + (w/\lambda)^N} + \frac{\lambda^{-2N}w^{N+M-2}}{1 + (w/\lambda)^N}.\end{aligned} \tag{4.60}$$

We can integrate (4.58), (4.59) and (4.60) with respect to dw, respectively, as

$$\begin{aligned}x_\sharp &= \lambda^{-N}\int_0^w \frac{w^{-2}}{1 + (w/\lambda)^N}dw \\ &\quad - \lambda^{-N}\int_0^w \frac{w^{M-2}}{1 + (w/\lambda)^N}dw + \text{Const.},\end{aligned} \tag{4.61}$$

$$\begin{aligned}x_\sharp &= -\lambda^{-N}w^{-1} - \lambda^{-2N}\int_0^w \frac{w^{N-2}}{1 + (w/\lambda)^N}dw \\ &\quad - \lambda^{-N}\int_0^w \frac{w^{M-2}}{1 + (w/\lambda)^N}dw + \text{Const.},\end{aligned} \tag{4.62}$$

$$\begin{aligned}x_\sharp &= \lambda^{-N}\int_0^w \frac{w^{-2}}{1 + (w/\lambda)^N}dw - \frac{\lambda^{-N}w^{M-1}}{M-1} \\ &\quad + \lambda^{-2N}\int_0^w \frac{w^{N+M-2}}{1 + (w/\lambda)^N}dw + \text{Const.}.\end{aligned} \tag{4.63}$$

Analogously, the three GHF-based solutions to be obtained from (4.61) to (4.63) should be identical. Therefore, equating any two of such three GHF-based solutions will lead to the formulation of another two recurrence formula between the two contiguous GHF.

The complete GHF-based solutions of (4.12) can be obtained by integrating the two integrals in (4.55), (4.56) and (4.57) for $\mid \lambda v \mid < 1$ plus those in (4.61), (4.62) and (4.63) for $\mid w/\lambda \mid < 1$ (or $\mid \lambda v \mid > 1$ after substituting $w = v^{-1}$ into $\mid w/\lambda \mid < 1$). Theoretically speaking, the GHF-based solutions of (4.12) or (4.14) should exclude a point where $\mid \lambda v \mid = 1$ (or $\mid w/\lambda \mid = 1$) because the GHF diverges at $\mid \lambda v \mid = 1$ (or $\mid w/\lambda \mid = 1$). Physically, such a point should have corresponded to the convergent limit of $x_\sharp$, where GVF profiles run parallel with the channel bottom, i.e., at $x_\sharp = \pm\infty$. However, contrary to the convergence criterion of the GHF on which (4.55), (4.56), (4.57), (4.61), (4.62) and (4.63) are so formulated, we have discovered that there exists no such a convergent limit for GVF profiles in adverse channels because the variable of the GHF, as will be shown below in each of such integrals, is of a negative value.

4.4.2 GHF-Based Solutions of GVF Profiles in Adverse Channels

(1) For $|\ \lambda v\ | < 1$ **:**

For $|\ \lambda v\ | < 1$, there are four different integrals in (4.61), (4.62) and (4.63). A general solution for these four integrals in terms of GHF can also be obtained with the help of the Mathematica software (or the Maple software) as

$$\int \frac{v^{\phi}}{1+(\lambda v)^N} dv = \frac{v^{\phi+1}}{\phi+1} g\left(\frac{\phi+1}{N}, -(\lambda v)^N\right) + \text{Const.}. \tag{4.64}$$

For the above mentioned four integrals in (4.61), (4.62) and (4.63), the parameter $\phi = N, 2N - M, N - M$, and $2N$, respectively. The GHF-based solutions of these four integrals are

$$\int_0^v \frac{v^N}{1+(\lambda v)^N} dv = \frac{v^{N+1}}{N+1} g\left(\frac{N+1}{N}, -(\lambda v)^N\right), \tag{4.65}$$

$$\int_0^v \frac{v^{2N-M}}{1+(\lambda v)^N} dv = \frac{v^{2N-M+1}}{2N-M+1} g\left(\frac{2N-M+1}{N}, -(\lambda v)^N\right), \tag{4.66}$$

$$\int_0^v \frac{v^{N-M}}{1+(\lambda v)^N} dv = \frac{v^{N-M+1}}{N-M+1} g\left(\frac{N-M+1}{N}, -(\lambda v)^N\right), \tag{4.67}$$

$$\int_0^v \frac{v^{2N}}{1+(\lambda v)^N} dv = \frac{v^{2N+1}}{2N+1} g\left(\frac{2N+1}{N}, -(\lambda v)^N\right). \tag{4.68}$$

Replacing the two integrals in each of (4.55), (4.56) and (4.57) using the GHF-based solutions listed in (4.65) to (4.68), we can express these three possible representations of the solution of (4.12) for $|\ \lambda v\ | < 1$ as

$$\begin{aligned} x_{\sharp} = &\frac{v^{N-M+1}}{N-M+1} g\left(\frac{N-M+1}{N}, -(\lambda v)^N\right) \\ &- \frac{v^{N+1}}{N+1} g\left(\frac{N+1}{N}, -(\lambda v)^N\right) + \text{Const.}, \end{aligned} \tag{4.69}$$

$$\begin{aligned} x_{\sharp} = &\frac{v^{N-M+1}}{N-M+1} - \frac{\lambda^N v^{2N-M+1}}{2N-M+1} g\left(\frac{2N-M+1}{N}, -(\lambda v)^N\right) \\ &- \frac{v^{N+1}}{N+1} g\left(\frac{N+1}{N}, -(\lambda v)^N\right) + \text{Const.}, \end{aligned} \tag{4.70}$$

$$x_{\sharp} = \frac{v^{N-M+1}}{N-M+1} g\left(\frac{N-M+1}{N}, -(\lambda v)^N\right) - \frac{v^{N+1}}{N+1} + \frac{\lambda^N v^{2N+1}}{2N+1} g\left(\frac{2N+1}{N}, -(\lambda v)^N\right) + \text{Const.}. \tag{4.71}$$

Because the three GHF-based solutions so obtained from (4.55) to (4.57), i.e., (4.69) to (4.71) are identical, comparing these GHF-based solutions yields the following two identities:

$$g\left(\frac{N+1}{N}, -(\lambda v)^N\right) = 1 - \frac{(N+1)(\lambda v)^N}{2N+1} g\left(\frac{2N+1}{N}, -(\lambda v)^N\right), \tag{4.72}$$

$$g\left(\frac{N-M+1}{N}, -(\lambda v)^N\right) = 1 - \frac{(N-M+1)(\lambda v)^N}{2N-M+1} g\left(\frac{2N-M+1}{N}, -(\lambda v)^N\right). \tag{4.73}$$

The above two equations are the recurrence formulas of two contiguous GHF of the variable, $-(\lambda v)^N$, with the composite parameters shown in (4.72) and (4.73). Equations (4.72) and (4.73) are also two special cases of the general identity relation of the two contiguous GHF for $|\ \lambda v\ | < 1$, as shown in Appendix B.

(2) For $|\ \lambda v\ | > 1$ **:**

As shown earlier, for $|\ \lambda v\ | > 1$ (or $|\ w/\lambda\ | < 1$ after substituting $v = w^{-1}$ into $|\ \lambda v\ | > 1$), there are four different integrals in (4.61) to (4.63). A general solution for these four integrals in terms of GHF can be obtained with the help of the Mathematica software (or the Maple software) as

$$\int \frac{w^{\phi}}{1+(w/\lambda)^N} dw = \frac{w^{\phi+1}}{\phi+1} g\left(\frac{\phi+1}{N}, -(w/\lambda)^N\right) + \text{Const.}. \tag{4.74}$$

The parameter $\phi = N-2$, $M-2$, -2, and $M+N-2$, respectively, for the above mentioned integrals in (4.61) to (4.63) yield the expressions

$$\int_0^w \frac{w^{N-2}}{1+(w/\lambda)^N} dw = \frac{w^{N-1}}{N-1} g\left(\frac{N-1}{N}, -(w/\lambda)^N\right), \tag{4.75}$$

$$\int_0^w \frac{w^{M-2}}{1+(w/\lambda)^N} dw = \frac{w^{M-1}}{M-1} g\left(\frac{M-1}{N}, -(w/\lambda)^N\right), \tag{4.76}$$

$$\int_0^w \frac{w^{-2}}{1+(w/\lambda)^N} dw = \frac{w^{-1}}{-1} g\left(\frac{-1}{N}, -(w/\lambda)^N\right), \tag{4.77}$$

$$\int_0^w \frac{w^{N+M-2}}{1+(w/\lambda)^N} dw = \frac{w^{N+M-1}}{N+M-1} g\left(\frac{N+M-1}{N}, -(w/\lambda)^N\right). \tag{4.78}$$

Replacing the two integral terms in each of (4.61), (4.62) and (4.63) using the GHF-based solutions listed in (4.75) to (4.78), we can express these three possible representations of the solution of (4.13) as

$$x_{\sharp} = -\lambda^{-N} w^{-1} g\left(\frac{-1}{N}, -(w/\lambda)^N\right) - \frac{\lambda^{-N} w^{M-1}}{M-1} g\left(\frac{M-1}{N}, -(w/\lambda)^N\right) + \text{Const.}, \tag{4.79}$$

$$x_{\sharp} = -\lambda^{-N} w^{-1} - \frac{\lambda^{-2N} w^{N-1}}{N-1} g\left(\frac{N-1}{N}, -(w/\lambda)^N\right) - \frac{\lambda^{-N} w^{M-1}}{M-1} g\left(\frac{M-1}{N}, -(w/\lambda)^N\right) + \text{Const.}, \tag{4.80}$$

$$x_{\sharp} = -\lambda^{-N} w^{-1} g\left(\frac{-1}{N}, -(w/\lambda)^N\right) - \frac{\lambda^{-N} w^{M-1}}{M-1} + \lambda^{-2N} \frac{w^{N+M-1}}{N+M-1} g\left(\frac{N+M-1}{N}, -(w/\lambda)^N\right) + \text{Const.}, \tag{4.81}$$

which should be valid in the domain of $| w/\lambda | < 1$. Because these three GHF-based solutions listed from (4.79) to (4.81) so obtained should be identical, comparing any two of them establishes the following two identities.

$$g\left(-\frac{1}{N}, -(w/\lambda)^N\right) = 1 + \frac{(w/\lambda)^N}{N-1} g\left(\frac{N-1}{N}, -(w/\lambda)^N\right), \tag{4.82}$$

$$g\left(\frac{M-1}{N}, -(w/\lambda)^N\right) = 1 - \frac{(M-1)(w/\lambda)^N}{N+M-1} g\left(\frac{N+M-1}{N}, -(w/\lambda)^N\right). \tag{4.83}$$

To express Eqs. (4.79) to (4.81) in terms of v, substituting $w = v^{-1}$ into the equations yields

$$x_{\sharp} = -\lambda^{-N} v g\left(-\frac{1}{N}, -(\lambda v)^{-N}\right) - \frac{\lambda^{-N} v^{-M+1}}{M-1} g\left(\frac{M-1}{N}, -(\lambda v)^{-N}\right) + \text{Const.}, \tag{4.84}$$

$$x_{\sharp} = -\lambda^{-N} v - \frac{\lambda^{-2N} v^{-N+1}}{N-1} g\left(\frac{N-1}{N}, -(\lambda v)^{-N}\right) - \frac{\lambda^{-N} v^{-M+1}}{M-1} g\left(\frac{M-1}{N}, -(\lambda v)^{-N}\right) + \text{Const.}, \tag{4.85}$$

$$x_{\sharp} = -\lambda^{-N} v g\left(-\frac{1}{N}, -(\lambda v)^{-N}\right) - \frac{\lambda^{-N} v^{-M+1}}{M-1} + \frac{\lambda^{-2N} v^{-N-M+1}}{N+M-1} g\left(\frac{N+M-1}{N}, -(\lambda v)^{-N}\right) + \text{Const.}. \tag{4.86}$$

The GHFs in Eqs. (4.84) to (4.86) must be valid in the domain $| \lambda v | > 1$. Because these three GHF-based solutions so obtained are identical, comparing any two of them establishes the following two identities.

$$g\left(-\frac{1}{N}, -(\lambda v)^{-N}\right) = 1 + \frac{(\lambda v)^{-N}}{N-1} g\left(\frac{N-1}{N}, -(\lambda v)^{-N}\right), \tag{4.87}$$

$$g\left(\frac{M-1}{N}, -(\lambda v)^{-N}\right) = 1 - \frac{(M-1)(\lambda v)^{-N}}{N+M-1} g\left(\frac{N+M-1}{N}, -(\lambda v)^{-N}\right). \tag{4.88}$$

It is noted that (4.82) and (4.83) are identical to (4.87) and (4.88), respectively, because they can be transformed to each other via $w = v^{-1}$. Also, these recurrence formulas are special cases of the general identity relation between the two contiguous GHF, as postulated in Appendix B.

Although the absolute value of the variable in the GHF, i.e., $| \lambda v |$, has been imposed to derive the GHF-based solutions from both (4.12) and (4.14), the complete GHF-based solutions of (4.12) for GVF profiles in adverse channels can only cover the physically possible domain of $| \lambda v | \leq 0$. Besides, because the six GHF-based solutions of (4.12), i.e., (4.69) to (4.71) in the domain of $| \lambda v | < 1$ as well as (4.84) to (4.86) in the domain of $| \lambda v | > 1$, are valid for $\lambda v \geq 0$ in reality, such six GHF-based solutions, if plotted, are indeed identical, provided that a boundary condition prescribed in the evaluation of the "Const." for each of (4.69) [or (4.70), (4.71)] and (4.84) [or (4.85) and (4.86)] lies in the same domain of any of such six GHF-based solutions used in plotting the A profiles.

It is also found that the two GHF appearing in each of (4.69) [or (4.70), (4.71)] and (4.84) [or (4.85) and (4.86)] converge at $\lambda v = 1$ because the variable of both GHF is of a negative value through its entire domain of $\lambda v \geq 0$. It is thus inferred that the A1 profile does not exist, while any one of such six GHF-based solutions is only required to plot the A2 and A3 profiles because it makes no difference to opt among (4.69) [or (4.70), (4.71)] or among (4.84) [or (4.85) and (4.86)] due to their respective convertibility through recurrence formulas, or ditto between (4.69) [or (4.70), (4.71)] and (4.84) [or (4.85) and (4.86)] thanks to their transformability in virtue of (F.1) or (F.2), as verified in Appendix F. For convenience in prescribing

Table 4.1 Equations for h_c-based dimensionless GVF profiles using the GHF on all types of slopes except for the horizontal slope

Equations for dimensionless GVF profiles expressed by using the GHF and hydraulic exponents, M and N	Eq. no.
For H2 ($1 \le v < \infty$) and H3 ($0 \le v \le 1$):	
$$x_\sharp = \frac{v^{N-M+1}}{N-M+1} - \frac{v^{N+1}}{N+1} + \text{Const.}.$$	(4.11)
For M1 ($\lambda^{-1} < v < \infty$)[a], S1 ($1 \le v < \infty$) and S2 ($\lambda^{-1} < v \le 1$):	
$$x_\sharp = \lambda^{-N} v g\left(\frac{-1}{N}, (\lambda v)^{-N}\right) + \frac{\lambda^{-N} v^{-M+1}}{M-1} g\left(\frac{M-1}{N}, (\lambda v)^{-N}\right) + \text{Const.}.$$	(4.47)
For M2 ($1 \le v < \lambda^{-1}$), M3 ($0 \le v \le 1$) and S3 ($0 \le v < \lambda^{-1}$):	
$$x_\sharp = \frac{v^{N-M+1}}{N-M+1} g\left(\frac{N-M+1}{N}, (\lambda v)^N\right) - \frac{v^{N+1}}{N+1} g\left(\frac{N+1}{N}, (\lambda v)^N\right) + \text{Const.}.$$	(4.32)
For C1 ($1 < v < \infty$) with $\lambda = 1$:	
$$x_\sharp = v g\left(\frac{-1}{N}, v^{-N}\right) + \frac{v^{-M+1}}{M-1} g\left(\frac{M-1}{N}, v^{-N}\right) + \text{Const.}.$$	(4.90)
For C3 ($0 \le v < 1$) with $\lambda = 1$:	
$$x_\sharp = \frac{v^{N-M+1}}{N-M+1} g\left(\frac{N-M+1}{N}, v^N\right) - \frac{v^{N+1}}{N+1} g\left(\frac{N+1}{N}, v^N\right) + \text{Const.}.$$	(4.89)
For A2 ($1 \le v < \infty$) and A3 ($0 \le v \le 1$):	
$$x_\sharp = \frac{v^{N-M+1}}{N-M+1} g\left(\frac{N-M+1}{N}, -(\lambda v)^N\right) - \frac{v^{N+1}}{N+1} g\left(\frac{N+1}{N}, -(\lambda v)^N\right) + \text{Const.}.$$	(4.69)

[a]The variable $v = h/h_c$ and the parameter $\lambda = h_c/h_n$. h_n is not real for flow in adverse channels

a boundary condition within the domain of $0 \le \lambda v < \lambda$ (or $0 \le v < 1$) that can accurately compute the A3 profile as well as for facilitating the conversion between the h_n-based solution (i.e., to be shown later) and the h_c-based solution, we select (4.69) as typical of the complete solutions of (4.12) spanning $0 \le \lambda v < \infty$ (or $0 \le v < \infty$) unless a circumstance in favor of (4.69) [or (4.70), (4.71)], or (4.84) [or (4.85) and (4.86)] arises otherwise. Table 4.1 shows the summary of equations for the h_c-based dimensionless GVF profiles expressed by using the GHF for flow in sustaining channels and adverse channels. Table 4.2 shows the summary of the GHF-based solutions of GVF profiles for the special case of $M = 3$ and $N = 10/3$.

Table 4.2 Equations for h_c-based dimensionless GVF profiles with $M = 3$ and $N = 10/3$

Equations for dimensionless GVF profiles expressed by using the GHF and $M = 3$ and $N = 10/3$	Eq. no.
For H2 ($1 \leq v < \infty$) and H3 ($0 \leq v \leq 1$):	
$x_\sharp = \frac{3}{4}v^{\frac{4}{3}} - \frac{3}{13}v^{\frac{13}{3}} + \text{Const.}.$	(4.91)
For M1 ($\lambda^{-1} < v < \infty$)[a], S1 ($1 \leq v < \infty$) and S2 ($\lambda^{-1} < v \leq 1$):	
$x_\sharp = \lambda^{-\frac{10}{3}}vg\left(\frac{-3}{10}, (\lambda v)^{-\frac{10}{3}}\right) + \frac{1}{2}\lambda^{-\frac{10}{3}}v^{-2}g\left(\frac{3}{5}, (\lambda v)^{-\frac{10}{3}}\right) + \text{Const.}.$	(4.92)
For M2 ($1 \leq v < \lambda^{-1}$), M3 ($0 \leq v \leq 1$) and S3 ($0 \leq v < \lambda^{-1}$):	
$x_\sharp = \frac{3}{4}v^{\frac{4}{3}}g\left(\frac{2}{5}, (\lambda v)^{\frac{10}{3}}\right) - \frac{3}{13}v^{\frac{13}{3}}g\left(\frac{13}{10}, (\lambda v)^{\frac{10}{3}}\right) + \text{Const.}.$	(4.93)
For C1 ($1 < v < \infty$) with $\lambda = 1$:	
$x_\sharp = vg\left(\frac{-3}{10}, v^{-\frac{10}{3}}\right) + \frac{1}{2}v^{-2}g\left(\frac{3}{5}, v^{-\frac{10}{3}}\right) + \text{Const.}.$	(4.94)
For C3 ($0 \leq v < 1$) with $\lambda = 1$:	
$x_\sharp = \frac{3}{4}v^{\frac{4}{3}}g\left(\frac{2}{5}, v^{\frac{10}{3}}\right) - \frac{3}{13}v^{\frac{13}{3}}g\left(\frac{13}{10}, v^{\frac{10}{3}}\right) + \text{Const.}.$	(4.95)
For A2 ($1 \leq v < \infty$) and A3 ($0 \leq v \leq 1$):	
$x_\sharp = \frac{3}{4}v^{\frac{4}{3}}g\left(\frac{2}{5}, -(\lambda v)^{\frac{10}{3}}\right) - \frac{3}{13}v^{\frac{13}{3}}g\left(\frac{13}{10}, -(\lambda v)^{\frac{10}{3}}\right) + \text{Const.}.$	(4.96)

[a]The variable $v = h/h_c$ and the parameter $\lambda = h_c/h_n$. h_n is not real for flow in adverse channels

4.4.3 Alternative Method to Get the GVF Profiles in Adverse Channels

It is worthwhile to iterate that we can alternatively get the solution (4.84) or [(4.85) and (4.86)] through an established relation connecting one GHF in the domain of $\mid z \mid < 1$ to two GHF in the domain of $\mid z \mid > 1$ such as (F.1) or (F.2) in Appendix F. Undoubtedly, such a relation enables one to transform the GHF-based solution of (4.12) in the domain of $\mid \lambda v \mid < 1$, i.e., (4.69) [or (4.70) and (4.71)], to the GHF-based solution of (4.12) in the domain of $\mid \lambda v \mid > 1$, i.e., (4.84) [or (4.85) and (4.86)]. Although we have thus far obtained the GHF-based solution of (4.12) in the domain of $\mid \lambda v \mid > 1$ from (4.14) [i.e., the transformed form of (4.12) with its variable, w, being expressed as v^{-1}] rather than through (4.12), we can prove that (4.84) [or (4.85) and (4.86)] is also obtainable from (4.69) [or (4.70) and (4.71)] using the two relations, i.e. (F.6) and (F.7) in Appendix F, to connect one GHF in the domain of $\mid \lambda v \mid < 1$ to two GHF in the domain of $\mid \lambda v \mid > 1$. It is proved in Appendix F that (F.8) so transformed from (4.69) through (F.6) and (F.7) is identical to (4.84). Using similar transformation processes, we can also obtain (4.85) and (4.86) directly from (4.70) and (4.71), and vice versa.

4.5 Summary

1. In this chapter, we have successfully established the h_c-based dimensionless GVF profiles expressed in terms of GHF for flow in sustaining channels as well as in adverse channels. The complete GHF-based solutions of (4.4) for the h_c-based dimensionless one-dimensional GVF profiles in sustaining channels should consist of any combination of (4.32) to (4.34) for $0 \leq (h_c/h_n)v < 1$ with (4.47) to (4.49) for $(h_c/h_n)v > 1$. By implication, any of such nine combinations can represent the complete GHF-based solutions of (4.4) for $0 \leq (h_c/h_n)v < \infty$ except at $(h_c/h_n)v = 1$, where $x_\sharp = \pm\infty$. Any difference in the GHF-expressions among them can be reconciled through transformation using the recurrence formulas, such as (4.35), (4.36), (4.50) and (4.51). The general form of the recurrence formula is shown in (3.34) and (B.4) in Appendix B. For comparison with the h_n-based dimensionless GVF profiles plotted in Chap. 3, we select the combination of (4.32) and (4.47) as typical of the complete solutions of (4.4) spanning $0 \leq (h_c/h_n)v < 1$ and $(h_c/h_n)v > 1$, respectively, unless a circumstance in favor of the other combinations arises otherwise.
2. Although we have obtained the GHF-based solutions of (4.4) in the domain of $\mid (h_c/h_n)v \mid > 1$ from (4.13) [i.e., a transformed form of (4.4) with its variable, w, being expressed as v^{-1}] rather than through (4.4), we can prove that (4.47) is also obtainable from (4.32) using the two relations, i.e. (F.3) and (F.4), to connect one GHF in the domain of $\mid (h_c/h_n)v \mid < 1$ to two GHF in the domain of $\mid (h_c/h_n)v \mid > 1$. It is proved in Appendix F that (F.5) so transformed from (4.32) through (F.3) and (F.4) is identical to (4.47).
3. The particular solutions of (4.32) and (4.37) for $\lambda = 1$ can be written as

$$x_\sharp = \frac{v^{N-M+1}}{N-M+1} g\left(\frac{N-M+1}{N}, v^N\right) - \frac{v^{N+1}}{N+1} g\left(\frac{N+1}{N}, v^N\right) + \text{Const.}, \tag{4.89}$$

$$x_\sharp = vg\left(\frac{-1}{N}, v^N\right) + \frac{v^{-M+1}}{M-1} g\left(\frac{M-1}{N}, v^{-N}\right) + \text{Const.}. \tag{4.90}$$

Equations (4.89) and (4.90) are specifically referred to as the solutions for h_c-based C3 profiles ($0 \leq v < 1$) and C1 profiles ($1 < v < \infty$), respectively.
4. We have obtained the GHF-based solutions of (4.12) for the h_c-based dimensionless GVF profiles in adverse channels in the domain of $\mid (h_c/h_n)v \mid > 1$ from (4.14) [i.e., a transformed form of (4.12) with its variable, w, being expressed as v^{-1}] rather than through (4.12), but we can prove that (4.84) is also obtainable from (4.69) using two relations, i.e. (F.6) and (F.7), to connect one GHF in the domain of $\mid (h_c/h_n)v \mid < 1$ to two GHF in the domain of $\mid (h_c/h_n)v \mid > 1$. It is proved in Appendix F that (F.8) so transformed from (4.69) through (F.6) and (F.7) is identical to (4.84).

5. The complete GHF-based solutions of (4.12) are found to cover only the physically possible domain of $(h_c/h_n)v \geq 0$. Because the six GHF-based solutions of (4.12), i.e., (4.69) to (4.71) in the domain of $|\ (h_c/h_n)v\ | < 1$ as well as (4.84) to (4.86) in the domain of $|\ (h_c/h_n)v\ | > 1$ are all valid for $(h_c/h_n)v \geq 0$ in reality, such six GHF-based solutions, if plotted, are indeed identical, provided that a boundary condition prescribed in the evaluation of the "Const." for each of (4.69) to (4.71) and (4.84) to (4.86) lies in the same domain of any of such GHF-based solutions used in plotting the A profiles. It is thus inferred that the A1 profile does not exist, while one needs only any of such six GHF-based solutions in plotting the A2 and A3 profiles.
6. It is found that (4.32) and (4.47) span the domains of $0 \leq (h_c/h_n)v < 1$ and $(h_c/h_n)v > 1$, respectively. We have chosen them as a set of the nine possible combinations of the GHF-based solutions of (4.4) for the h_c-based dimensionless GVF profiles in sustaining channels, while (4.69) covering the domain of $(h_c/h_n)v \geq 0$ can be deemed as one of the six GHF-based solutions of (4.12) for the h_c-based dimensionless GVF profiles in adverse channels. Table 4.1 shows the summary of equations for the h_c-based dimensionless GVF profiles expressed by using the GHF for flow in sustaining channels and adverse channels. Table 4.2 shows the summary of the GHF-based solutions of GVF profiles for the special case of $M = 3$ and $N = 10/3$. A more in-depth analysis of the properties of the GHF-based GVF profiles obtained herein will be presented in the next chapter.

References

Abdulrahman A (2010) Limit slope in uniform flow computations: Direct solution for rectangular channels. J Hydrol Eng 15(10):808–812

Allen J, Enever KJ (1968) Water surface profiles in gradually varied open-channel flow. Proc Inst Civil Eng 41:783–811

Bakhmeteff BA (1932) Hydraulics of open channels. McGraw-Hill, New York

Bresse J (1860) Cours de mécanique appliqué. hydraulique. Mallet-Bachelier, Paris, 2e partie edition

Chen CL (1991) Unified theory on power laws for flow resistance. J Hydraul Eng ASCE 117(3): 371–389

Chen CL, Wang CT (1969) Nondimensional gradually varied flow profiles. J Hydraul Div ASCE 95(HY5):1671–1686

Chow VT (1955) Integrating the equation of gradually varied flow. Proceedings Paper No.838, ASCE (81):1–32

Chow VT (1957) Closure of discussions on integrating the equation of gradually varied flow. J Hydraul Div ASCE 83(1):9–22

Chow VT (1959) Open-channel hydraulics. McGraw-Hill, New York

French R (1986) Open-channel hydraullics. McGraw-Hill Book Company, New York

Jan CD, Chen CL (2012) Use of the Gaussian hypergeometric function to solve the equation of gradually-varied flow. J Hydrol 456–457:139–145

Jan CD, Chen CL (2013) Gradually varied open-channel flow profiles normalized by critical depth and analytically solved by using gaussian hypergeometric functions. Hydrol Earth Syst Sci (HESS) 17:973–987

Kumar A (1978) Integral solutions of the gradually varied equation for rectangular and triangular channels. Proc Inst Civil Eng 65(2):509–515

Kumar A (1979) Gradually varied surface profiles in horizontal and adversely sloping channels. Proc Inst Civil Eng Part 2(67):435–452

Matzke AE (1937) Varied flow in open channels of adverse slope. Trans ASCE 102:651–660

Rouse H (1965) Critical analysis of open-channel resistance. J Hydraul Div ASCE 91(HY4):1–25

Subramanya K (2009) Flow in open channels, 3rd edn. McGraw-Hill, Singapore

Wolfram S (1996) The mathematica book, 3rd edn. Wolfram Media and Cambridge University Press, Champaign

Zaghloul NA (1990) A computer model to calculate varied flow functions for circular channels. Adv Eng Softw 12(3):106–122

Zaghloul NA (1992) Gradually varied flow in circular channels with variable roughness. Adv Eng Softw 15(1):33–42

Chapter 5
Analysis of the GHF-Based Solutions of h_c-Based GVF Profiles

5.1 Introduction

The equation of one-dimensional gradually-varied flow (GVF) in sustaining and non-sustaining open channels is normalized using the critical depth, h_c, and then solved by the direct integration method with the use of the Gaussian hypergeometric functions (GHF), as shown in Chap. 4. This chapter presents the properties of the GHF-based solutions of the h_c-based GVF profiles obtained in Chap. 4. The h_c-based GVF solutions expressed in terms of the GHF have been proved to be more useful and versatile than its counterpart normalized using the normal depth, h_n, because the h_c-based dimensionless mild (M) and adverse (A) profiles can asymptotically reduce to the h_c-based dimensionless horizontal (H) profiles as $h_c/h_n \rightarrow 0$. Nevertheless, it cannot prevail in case of the h_n-based dimensionless M and A profiles. This chapter conducts an in-depth analysis on the effects of h_c/h_n and a function parameter of the GHF (i.e., N, one of the hydraulic exponents, M and N) on the h_c-based dimensionless GVF profiles, their curvature and inflection points, and the existence of singularities. The results show that the features and properties of the h_c-based dimensionless GVF profiles expressed in terms of the GHF reflect the very same trend of their equivalents based on h_n for flow in sustaining channels. We can also prove that the asymptotic reduction of the sole inflection point on the M or A profile to their counterpart on the H profile as $h_c/h_n \rightarrow 0$ is possible only by taking the limit of their respective h_c-based dimensionless conditions for the existence of an inflection point at $h_c/h_n = 0$. However, their equivalents based on h_n prove otherwise. It is found that the acceptance of any real numbers by the function parameters (M and N) of the GHF-based solution can relax restrictions placed on the assumptions of the constant M- and N-values in the direct integration of the two indefinite integrals appearing in the GVF equation.

C.-D. Jan, *Gradually-varied Flow Profiles in Open Channels*,
Advances in Geophysical and Environmental Mechanics and Mathematics,
DOI: 10.1007/978-3-642-35242-3_5, © Springer-Verlag Berlin Heidelberg 2014

5.2 Analysis of the GHF-Based Solutions of Equations (4.4) and (4.12)

5.2.1 Plotting the GHF-Based Solutions on the $(x_\sharp, v)$-Plane

It is worthwhile to reiterate that (4.32) and (4.47), which span, respectively, the domains of $0 \leq \lambda v < 1$ and $\lambda v > 1$, have been chosen as a set of the nine possible combinations of the GHF-based solutions of (4.4) for h_c-based dimensionless GVF profiles in sustaining channels, while (4.69) covering the domain of $\lambda v \geq 0$ has been used as one of the six equivalent GHF-based solutions of (4.12) for h_c-based dimensionless GVF profiles in adverse channels. A plot of (4.32), (4.47) and (4.69) on the $(x_\sharp, v)$-plane with λ $(= h_c/h_n)$ as a parameter may help gain an insight into the uniqueness and versatility of such GHF-based solutions.

Upon prescription of the boundary conditions, one for each solution, we can determine the constants of integration, "Const.", in (4.32), (4.47), and (4.69), thereby plotting these equations on the $(x_\sharp, v)$-plane with λ as a parameter, as shown in Fig. 5.1. In this figure, three boundary conditions are arbitrarily prescribed at $(x_\sharp, v) = (0, 0)$, $(10, 10)$, and $(0, 0)$ for (4.32), (4.47), and (4.69), respectively. Any N-value computed from $N = 2m+3$ for fully rough flows or $N = 3$ for hydraulically smooth flows in wide channels ($M = 3$) may be used. However, for comparison, the same N-value used in plotting the h_n-based solutions, as shown in Fig. 3.1, can be selected in plotting the h_c-based solutions. We thus use $M = 3$ and $N = 10/3$ (i.e., a typical value corresponding to the Manning formula) as an example to plot the h_c-based solution curves on the $(x_\sharp, v)$-plane herein. The GHF-based solutions of the h_c-based GVF equation under the conditions of $M = 3$ and $N = 10/3$ are also shown in Table 4.2.

5.2.2 Mild (M), Steep (S), and Adverse (A) Profiles

Examination of Fig. 5.1 reveals that two solution curves drawn for the critical value of λ (i.e., $\lambda = h_c/h_n = 1$) and the asymptotic value of λ [i.e., λ $(= h_c/h_n) = 0$] combined with the horizontal line at $v = 1$ and the horizontal asymptotes at $\lambda v = 1$ (or $v = h_n/h_c$) can divide the entire domain of the solution curves on the $(x_\sharp, v)$-plane into nine regions, which are constituted by a combination of three slopes and three zones, i.e., the same division also adopted by Chow (1959). The three slopes consist of mild (M) slopes ($0 < h_c/h_n < 1$), steep (S) slopes ($1 < h_c/h_n < \infty$), and adverse (A) slopes ($0 < h_c/h_n < \infty$, in which h_n is fictitious, as defined earlier), while the three zones are composed of zone 1 ($h_n/h_c < v < \infty$ for M profiles or $1 \leq v < \infty$ for S profiles), zone 2 ($1 \leq v < h_n/h_c$ for M profiles, $h_n/h_c < v \leq 1$ for S profiles, or $1 \leq v < \infty$ for A profiles), and zone 3 ($0 \leq v \leq 1$ for M profiles, $0 \leq v < h_n/h_c$ for S profiles, or $0 \leq v \leq 1$ for A profiles). Apparently, the particular solution curves drawn by (4.32) and (4.47) with $h_c/h_n = 1$ separate the region of M profiles from

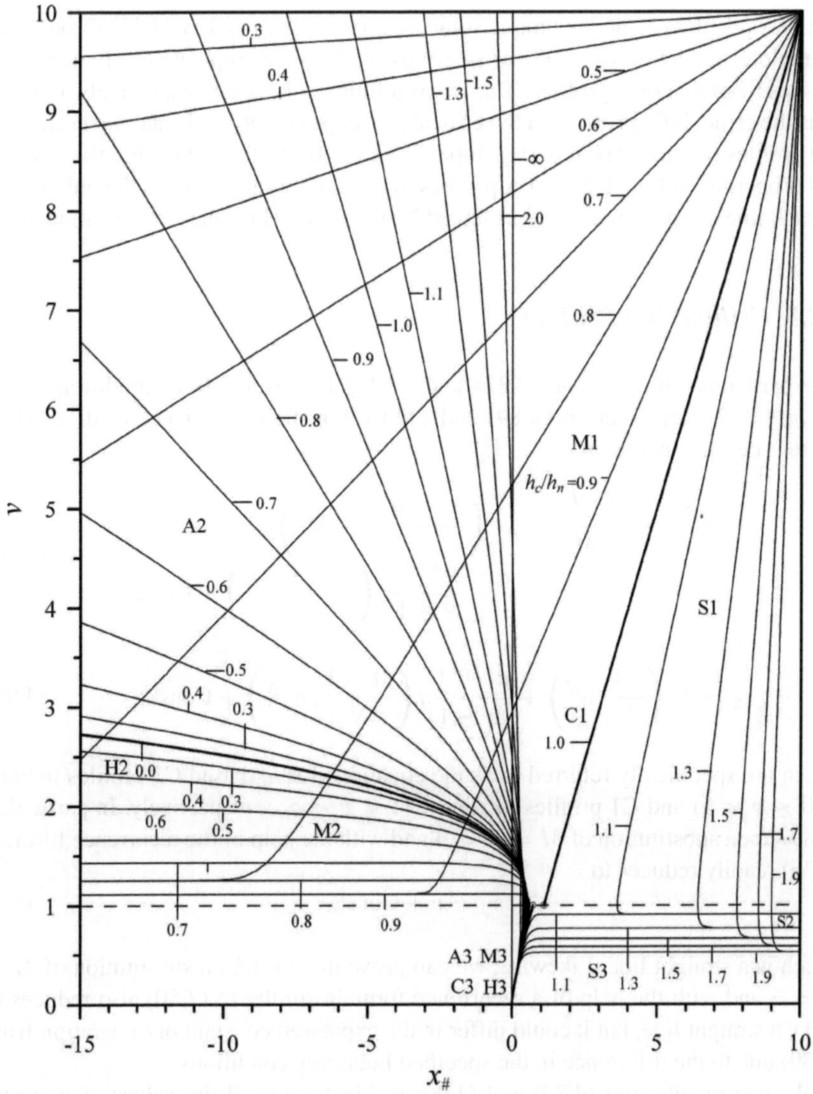

Fig. 5.1 Plot of the h_c-based dimensionless GVF profiles for $M = 3$ and $N = 10/3$ on the $(x_\sharp, v)$-plane with h_c/h_n as a parameter. For discerning the woven profiles from one another, the boundary conditions of all the profiles plotted for (4.47) and (4.90) in the domain of $v > h_n/h_c$ as well as plotted for (4.32) and (4.89) in the domain of $0 \le v < h_n/h_c$, for (4.11) in the domain of $0 \le v < \infty$, and for (4.69) in the domain of $0 \le v < \infty$ are arbitrarily, judiciously set at $(x_\sharp, v) = (10, 10)$ and $(0, 0)$, respectively. As $h_c/h_n \to 0$, the solution curves plotted for (4.32) in the region of M2 and M3 profiles ($0 < h_c/h_n < 1$) as well as (4.69) in the region of A2 and A3 profiles ($0 < h_c/h_n < \infty$, in which h_n is fictitious reduce asymptotically to (4.11), i.e., the equation of the H2 and H3 profiles. There are two points on each profile to which the slope of the profile is infinite, one at $v = 0$ and the other at $v = 1$. *Two small open circles* shown in the graph at $(x_\sharp, v) = (1.09317, 1)$ and $(0.717658, 1)$ stand for two singularities on the slopes of the C1 and C3 profiles at $v = 1$, respectively. Jan and Chen (2013), reproduced with permission

that of S profiles, while the unique solution curve constructed by (4.11) through the asymptotic reduction of (4.32) or (4.47) as $h_c/h_n \to 0$ separate the region of M profiles from that of A profiles. These two solution curves are, respectively, referred to as the critical (C) profiles on the critical (C) slope ($h_c/h_n = 1$) and the horizontal (H) profiles on the horizontal (H) slope ($h_c/h_n = 0$). Before analyzing the features and properties of the M, S, and A profiles, we first focus attention on the uniqueness of such lines of demarcation represented by the C and H profiles in the following.

5.2.3 *Critical (C) Profiles*

The particular solutions of (4.32) and (4.47) for critical-flow condition ($\lambda = h_c/h_n = 1$), are given in (4.89) and (4.90). For the convenience of discussion, we rewrite them herein as

$$x_\sharp = \frac{v^{N-M+1}}{N-M+1} g\left(\frac{N-M+1}{N}, v^N\right) - \frac{v^{N+1}}{N+1} g\left(\frac{N+1}{N}, v^N\right) + \text{Const.}, \tag{4.89}$$

$$x_\sharp = vg\left(\frac{-1}{N}, v^N\right) + \frac{v^{-M+1}}{M-1} g\left(\frac{M-1}{N}, v^{-N}\right) + \text{Const.}, \tag{4.90}$$

which are specifically referred to as the equations for h_c-based C3 profiles in zone 3 ($0 \le v < 1$) and C1 profiles in zone 1 ($1 < v < \infty$), respectively. In particular, (4.89) upon substitution of $M = N = 3$ and with the help of the recurrence formula (4.36) readily reduces to

$$x_\sharp = v + \text{Const.}, \tag{5.1}$$

which is a straight line. Likewise, we can prove that (4.90) on substitution of $M = N = 3$ and with the help of a recurrence formula similar to (4.50) also reduces to (5.1), a straight line, but it could differ in the expressed constant of integration from (4.89) due to the difference in the specified boundary conditions.

We can readily plot (4.89) and (4.90) in Fig. 5.1 for all the values of v except $v = 1$, where both equations are undefined due to the existence of a singularity. The existence of a singularity in (4.89) and (4.90) at $v = 1$ can be readily realized from (4.4), which upon substitution of $h_c/h_n = 1$ yields

$$\frac{dv}{dx_\sharp} = \frac{1 - v^N}{v^{N-M}(1 - v^M)}. \tag{5.2}$$

Equation (5.2) is the expression of the slopes of GVF profiles on the C slope. The right-hand side of (5.2) is a rational function, whose numerator and denominator are zero at $v = 1$. Therefore, (4.89), (4.90), and (5.2) are all undefined at $v = 1$, connoting the existence of a singularity at $v = 1$. Two small open circles are used to mark the singularities in Fig. 5.1. However, if $M = N = 3$, there is no singularity at $v = 1$ because (5.2) can be readily reduced to $dv/dx_{\sharp} = 1$, whose right-hand side is no longer a rational function. It can be directly integrated into the form of (5.1), i.e., a straight line, which has been derived earlier either from (4.89) or (4.90). It is self-evident from Fig. 5.1 that both equations so plotted for $h_c/h_n = 1$ irrespective of the existence of a singularity at $v = 1$ are the lines of demarcation, which divide the domain of solutions curves obtained from (4.32) and (4.47) into two regions, one for M profiles ($0 < h_c/h_n \leq 1$) and the other for S profiles ($h_c/h_n \geq 1$).

5.2.4 Horizontal (H) Profiles in Zones 2 and 3

To separate the region of M profiles from that of A profiles is a solution curve plotted by using (4.11), to which (4.32) and (4.47) reduce asymptotically as λ ($= h_c/h_n$) $\rightarrow$ 0 from either side of (4.11), as shown in Fig. 5.1. For clarity in highlighting this line of demarcation, (4.11), i.e., a solution curve for $h_c/h_n = 0$, we draw Fig. 5.2, a close-up of Fig. 5.1 around the commonly prescribed boundary condition of (4.11), (4.32), (4.47), and (4.89) at $(x_{\sharp}, v) = (0, 0)$. We prove that (4.11) can be obtained by the asymptotic reduction of (4.32) or (4.69) as $h_c/h_n \rightarrow 0$, using the GHF definition, as elaborated in Appendix G. Attention should be focused on the fact that not all of the solution curves describable by (4.32) can asymptotically reduce to an H2 profile in zone 2 ($1 \leq v < \infty$) or an H3 profile in zone 3 ($0 \leq v \leq 1$) as $h_c/h_n \rightarrow 0$ because the limits of the first and second terms on the right-hand side of (4.32) at $h_c/h_n = 0$ for some profiles, such as the C3 profile ($0 \leq v < 1$) plotted by (4.89) [or (4.32) upon substitution of $h_c/h_n = 1$] and the S3 profiles ($0 \leq v < h_n/h_c$) on the S slope ($1 < h_c/h_n < \infty$), do not exist. These results along with the A2 and A3 profiles plotted using (4.69), which can also reduce asymptotically to (4.11) as $h_c/h_n \rightarrow 0$, are summarized in Table 5.1. As displayed clearly in Fig. 5.2, the H2 profile divides perfectly between the regions of M2 profiles and A2 profiles; so does the H3 profile between the regions of M3 profiles and A3 profiles.

5.2.5 Horizontal Asymptotes at $v = h_n/h_c$ for Various h_n/h_c-values

In addition to the solution curves for $\lambda = h_c/h_n = 1$ [i.e., (4.89) and (4.90), or (5.1) if $M = N = 3$] and those for $h_c/h_n \rightarrow 0$ [i.e., (4.11)] which are used as lines of demarcation, we need the horizontal asymptotes at $(h_c/h_n)v = 1$ (or $v = h_n/h_c$) to delineate further the entire domain of the solution curves plotted for GVF on the sustaining slope, as shown in Fig. 5.1. It is already known that (4.32) and (4.47) diverge along the horizontal asymptotes. However, unlike the sole horizontal

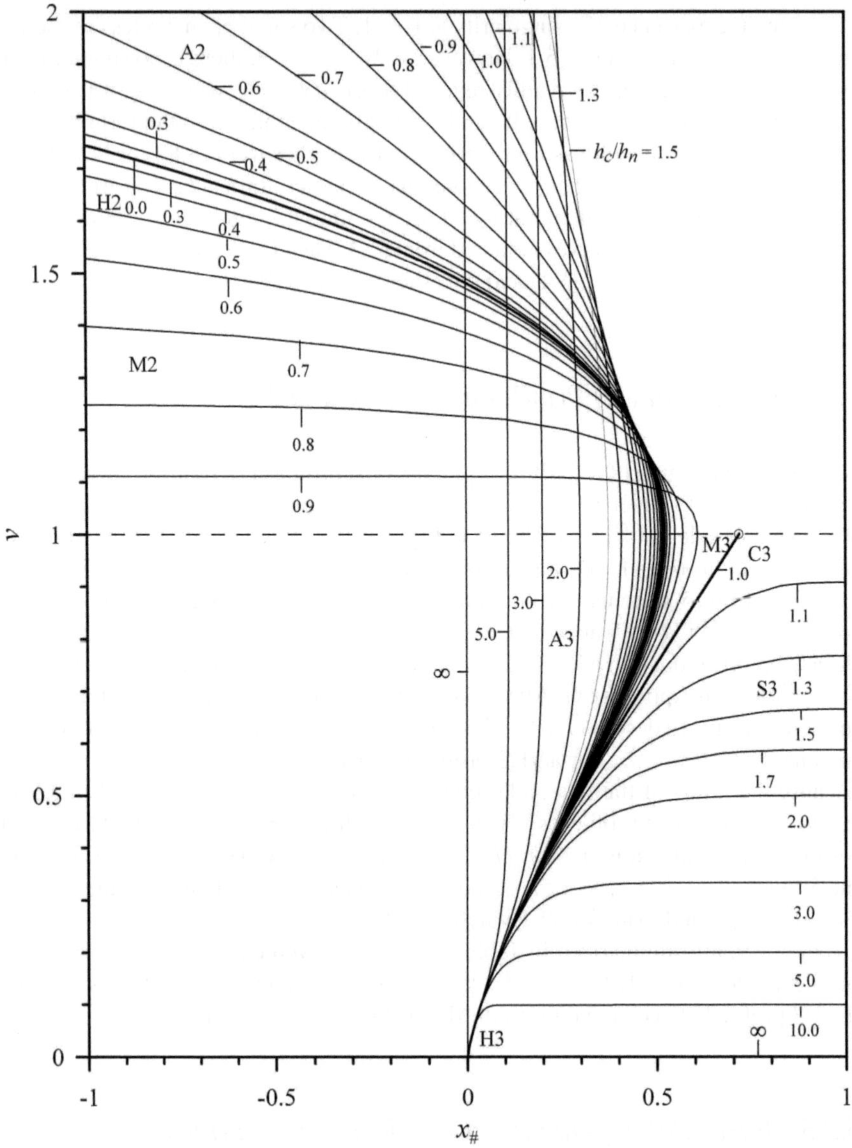

Fig. 5.2 A close-up of those H, M, C, S and A profiles which are plotted in Fig. 3.1 around the commonly prescribed boundary condition of (4.11), (4.32), (4.69), and (4.89) at $(x_\sharp, v) = (0, 0)$. This scale-up plot of such profiles reconfirms that the H2 profile divides perfectly between the regions of M2 profiles and A2 profiles; so does the H3 profile between the regions of M3 profiles and A3 profiles, as proved in Appendix B. It highlights a well-behaved pattern of GVF profiles in channels with θ varying from $\theta > 0$ [i.e., (4.32) in the domain of $0 \leq v < h_n/h_c$], passing through $\theta = 0$ [i.e., (4.11) in the domain of $0 \leq v < \infty$], and to $\theta < 0$ [i.e., (4.69) in the domain of $0 \leq v < \infty$], with emphasis on displaying the "evolving" variation of such solution curves with h_c/h_n. Jan and Chen (2013), reproduced with permission

Table 5.1 List of GVF profiles on the sustaining slope ($\theta > 0$) and adverse slope ($\theta < 0$) which can asymptotically reduce to the H2 or H3 profile on the horizontal slope ($\theta = 0$) as $\theta \to 0$ and their comparison with those which cannot

Before asymptotic reduction ($\theta > 0$ or $\theta < 0$)			After asymptotic reduction ($\theta = 0$)		
Types of GVF profiles	Valid domain of v	Equation applied (equation number)	Types of GVF profiles	Valid domain of v	Equation applied (equation number)
M1	$h_n/h_c < v < \infty$	(4.47)	N/A	$\infty < v$	None
M2	$1 \leq v < h_n/h_c$	(4.32)	H2	$1 \leq v < \infty$	(4.11)
M3	$0 \leq v \leq 1$	(4.32)	H3	$0 \leq v \leq 1$	(4.11)
C1	$1 < v < \infty$	(4.90)	N/A	$1 < v < \infty$	None
C3	$0 \leq v < 1$	(4.89)	N/A	$0 \leq v < 1$	None
S1	$1 \leq v < \infty$	(4.47)	N/A	$\infty < v$	None
S2	$h_n/h_c < v \leq 1$	(4.47)	N/A	$\infty < v$	None
S3	$0 \leq v < h_n/h_c$	(4.32)	N/A	$0 \leq v < \infty$	None
A2	$1 \leq v < \infty$	(4.69)	H2	$1 \leq v < \infty$	(4.11)
A3	$0 \leq v \leq 1$	(4.69)	H3	$0 \leq v \leq 1$	(4.11)

asymptote at $u = 1$ on the (x_*, u)-plane to represent the water surface running parallel with the channel bed as $x_* \to \pm\infty$ as shown in Fig. 3.1, the horizontal asymptotes at $v = h_n/h_c$ for various h_n/h_c-values on the $(x_\sharp, v)$-plane translate vertically. Accordingly, the horizontal asymptotes at $v = h_n/h_c$ on the $(x_\sharp, v)$-plane for the given h_n/h_c-values may be deemed as "movable" lines of demarcation drawn to divide the domain of v into two major regions: One region spans $0 \leq (h_c/h_n) < 1$ (or $0 \leq v < h_n/h_c$), which covers both zones 2 and 3 for M profiles as well as zone 3 for S profiles, and the other region extends over $(h_c/h_n)v > 1$ (or $v > h_n/h_c$), which covers zone 1 for M profiles as well as both zones 1 and 2 for S profiles.

In comparison, there exist no horizontal asymptotes at $v = h_n/h_c$ for various h_n/h_c-values in the domain of solution curves plotted for (4.69) on the $(x_\sharp, v)$-plane, which represent the h_c-based dimensionless GVF profiles on the adverse slope. Inspection of Fig. 5.1 clearly reveals that the solution curves plotted for (4.32) and (4.47) diverge at $v = h_n/h_c$ on the $(x_\sharp, v)$-plane, whereas those which are plotted for (4.69) do not diverge at $v = h_n/h_c$ for any h_n/h_c-value shown. The latter result may be attributed to the fact that h_n adopted in the derivation of (4.69) is simply fictitious (i.e., nonexistent).

Before classifying the solution curves so plotted for (4.32), (4.47), and (4.69), one may raise a question as to the uniqueness of a pattern of solution curves so plotted because the pattern can be affected by the boundary conditions and the values of M and N. For example, a change in the boundary value of $x_\sharp$ could have translated all the solution curves horizontally in response to the exact amount of the changed $x_\sharp$ as long as the other boundary value of v is kept intact. Nevertheless, the pattern of solution curves so plotted can also vary with the N-value on top of the boundary conditions even though the M-value is unchanged (i.e., $M = 3$ for flow in wide channels). In Fig. 5.1, we only prescribe one boundary condition for each of (4.32), (4.47), and (4.69), which are then substituted by $M = 3$ and $N = 10/3$ to plot the h_c-based dimensionless GVF profiles of a fully rough one-sixth power-law flow in wide sustaining and adverse channels with a full θ-range representing various h_c/h_n-values, including $h_c/h_n = 0$ and 1 corresponding to GVF profiles on the horizontal and critical slopes, respectively, as well as $h_c/h_n = \infty$ for both S and A profiles. Therefore, aside from the assigned values of M and N, we may claim that the pattern of solution curves so plotted is unique as long as there are no changes in the prescribed boundary conditions.

5.3 Classification of GHF-Based Solutions Using h_c/h_n

In contrast to eight types of the h_n-based GVF profiles, as classified in Chap. 3, there exist twelve types of the h_c-based GVF profiles, two on the H slope ($h_c/h_n = 0$), three on the M slope ($0 < h_c/h_n < 1$), two on the C slope ($h_c/h_n = 1$), three on the S slope ($1 < h_c/h_n < \infty$), and two on the A slope ($0 \leq h_c/h_n < \infty$, in which h_n is fictitious). The twelve h_c-based GVF profiles in three zones may, respectively, be referred to as H2 and H3; M1, M2, and M3; C1 and C3; S1, S2, and S3; A2 and A3. The twelve profiles so classified are the same as those classified by Chow (1959) except for C2, which is excluded from our classification because it is a singularity for all the N-values except for $N = 3$, as proved in Chap. 3. Such twelve profiles are classified by use of h_n/h_c (or its reciprocal, h_c/h_n) to demarcate appropriately the entire domain of v (i.e., $0 \leq v < \infty$) and then described using one of (4.11), (4.32), (4.47), and (4.69), as tabulated in Table 4.1. However, to be more specific, the two h_c-based GVF profiles on the C slope ($h_c/h_n = 1$) in Table 4.1 are described using (4.89) and (4.90) instead of (4.32) and (4.47) on substitution of $h_c/h_n = 1$, respectively.

Although the parameter, N, in the GHF-based solutions, as tabulated in Table 4.1, can be valid for any real number, we are presently interested in using the five N-values, 3, 10/3, 17/5, 7/2, and 11/3 for fully rough flows unless specified otherwise for other applications. As explained earlier, because it is always $N = 3$ for hydraulically smooth flows, the five N-values so chosen are determined from $N = 2m + 3$ corresponding to $m = 0$, 1/6, 1/5, 1/4, and 1/3, respectively, for fully rough flows.

Strictly speaking, we only need three boundary conditions, one for each of (4.32), (4.47), and (4.69) [or (4.89) and (4.90) if λ $(= h_c/h_n) = 1$], to determine "Const." in each equation, as stated previously. However, we may purposely use as many boundary conditions as needed for the three zones, to evaluate "Const." in each of (4.11), (4.32) [or (4.89) if $h_c/h_n = 1$], (4.47) [or (4.90) if $h_c/h_n = 1$], and (4.69). These equations on substitution of the respective values of "Const." yield the twelve particular GHF-based solutions of (4.4) and (4.12) for the h_c-based dimensionless GVF profiles, i.e., (5.3–5.14), as shown below.

(1) For H2 profile ($1 \le v < \infty$):

$$x_\sharp = x_{\sharp H2} + \frac{1}{N-M+1}(v^{N-M+1} - v_{H2}^{N-M+1}) - \frac{1}{N+1}(v^{N+1} - v_{H2}^{N+1}). \quad (5.3)$$

(2) For H3 profile ($0 \le v \le 1$):

$$x_\sharp = x_{\sharp H3} + \frac{1}{N-M+1}(v^{N-M+1} - v_{H3}^{N-M+1}) - \frac{1}{N+1}(v^{N+1} - v_{H3}^{N+1}). \quad (5.4)$$

(3) For M1 profile ($\lambda^{-1} \le v < \infty$):

$$\begin{aligned} x_\sharp = x_{\sharp M1} &+ \lambda^{-N}\left[vg\left(\frac{-1}{N}, (\lambda v)^{-N}\right) - v_{M1}g\left(\frac{-1}{N}, (\lambda v_{M1})^{-N}\right)\right] \\ &+ \frac{\lambda^{-N}}{M-1}\left[v^{-M+1}g\left(\frac{M-1}{N}, (\lambda v)^{-N}\right) - v_{M1}^{-M+1}g\left(\frac{M-1}{N}, (\lambda v_{M1})^{-N}\right)\right]. \end{aligned} \quad (5.5)$$

(4) For M2 profile ($1 \le v < \lambda^{-1}$):

$$\begin{aligned} x_\sharp = x_{\sharp M2} &+ \frac{v^{N-M+1}}{N-M+1}g\left(\frac{N-M+1}{N}, (\lambda v)^N\right) \\ &- \frac{v_{M2}^{N-M+1}}{N-M+1}g\left(\frac{N-M+1}{N}, (\lambda v_{M2})^N\right) \\ &- \frac{v^{N+1}}{N+1}g\left(\frac{N+1}{N}, (\lambda v)^N\right) + \frac{v_{M2}^{N+1}}{N+1}g\left(\frac{N+1}{N}, (\lambda v_{M2})^N\right). \end{aligned} \quad (5.6)$$

(5) For M3 profile ($0 \le v \le 1$):

$$\begin{aligned} x_\sharp = x_{\sharp M3} &+ \frac{v^{N-M+1}}{N-M+1}g\left(\frac{N-M+1}{N}, (\lambda v)^N\right) \\ &- \frac{v_{M3}^{N-M+1}}{N-M+1}g\left(\frac{N-M+1}{N}, (\lambda v_{M3})^N\right) \\ &- \frac{v^{N+1}}{N+1}g\left(\frac{N+1}{N}, (\lambda v)^N\right) + \frac{v_{M3}^{N+1}}{N+1}g\left(\frac{N+1}{N}, (\lambda v_{M3})^N\right). \end{aligned} \quad (5.7)$$

(6) For C1 profile ($1 < v < \infty$):

$$x_\sharp = x_{\sharp C1} + \left[vg\left(\frac{-1}{N}, v^{-N}\right) - v_{C1}g\left(\frac{-1}{N}, v_{C1}^{-N}\right)\right] + \frac{1}{M-1}\left[v^{-M+1}g\left(\frac{M-1}{N}, v^{-N}\right) - v_{C1}^{-M+1}g\left(\frac{M-1}{N}, v_{C1}^{-N}\right)\right]. \tag{5.8}$$

(7) For C3 profile ($0 \le v < 1$):

$$x_\sharp = x_{\sharp C3} + \frac{v^{N-M+1}}{N-M+1}g\left(\frac{N-M+1}{N}, v^N\right) - \frac{v_{C3}^{N-M+1}}{N-M+1}g\left(\frac{N-M+1}{N}, v_{C3}^N\right) - \frac{v^{N+1}}{N+1}g\left(\frac{N+1}{N}, v^N\right) + \frac{v_{C3}^{N+1}}{N+1}g\left(\frac{N+1}{N}, v_{C3}^N\right). \tag{5.9}$$

(8) For S1 profile ($1 \le v < \infty$):

$$x_\sharp = x_{\sharp S1} + \lambda^{-N}\left[vg\left(\frac{-1}{N}, (\lambda v)^{-N}\right) - v_{S1}g\left(\frac{-1}{N}, (\lambda v_{S1})^{-N}\right)\right] + \frac{\lambda^{-N}}{M-1}\left[v^{-M+1}g\left(\frac{M-1}{N}, (\lambda v)^{-N}\right) - v_{S1}^{-M+1}g\left(\frac{M-1}{N}, (\lambda v_{S1})^{-N}\right)\right]. \tag{5.10}$$

(9) For S2 profile ($\lambda^{-1} < v \le 1$):

$$x_\sharp = x_{\sharp S2} + \lambda^{-N}\left[vg\left(\frac{-1}{N}, (\lambda v)^{-N}\right) - v_{S2}g\left(\frac{-1}{N}, (\lambda v_{S2})^{-N}\right)\right] + \frac{\lambda^{-N}}{M-1}\left[v^{-M+1}g\left(\frac{M-1}{N}, (\lambda v)^{-N}\right) - v_{S2}^{-M+1}g\left(\frac{M-1}{N}, (\lambda v_{S2})^{-N}\right)\right]. \tag{5.11}$$

(10) For S3 profile ($0 \le v < \lambda^{-1}$):

$$x_\sharp = x_{\sharp S3} + \frac{v^{N-M+1}}{N-M+1}g\left(\frac{N-M+1}{N}, (\lambda v)^N\right) - \frac{v_{S3}^{N-M+1}}{N-M+1}g\left(\frac{N-M+1}{N}, (\lambda v_{S3})^N\right) - \frac{v^{N+1}}{N+1}g\left(\frac{N+1}{N}, (\lambda v)^N\right) + \frac{v_{S3}^{N+1}}{N+1}g\left(\frac{N+1}{N}, (\lambda v_{S3})^N\right). \tag{5.12}$$

(11) For A2 profile ($1 \le v < \infty$):

$$x_\sharp = x_{\sharp A2} + \frac{v^{N-M+1}}{N-M+1} g\left(\frac{N-M+1}{N}, -(\lambda v)^N\right) - \frac{v_{A2}^{N-M+1}}{N-M+1} g\left(\frac{N-M+1}{N}, -(\lambda v_{A2})^N\right) - \frac{v^{N+1}}{N+1} g\left(\frac{N+1}{N}, -(\lambda v)^N\right) + \frac{v_{A2}^{N+1}}{N+1} g\left(\frac{N+1}{N}, -(\lambda v_{A2})^N\right). \tag{5.13}$$

(12) For A3 profile ($0 < v \le 1$):

$$x_\sharp = x_{\sharp A3} + \frac{v^{N-M+1}}{N-M+1} g\left(\frac{N-M+1}{N}, -(\lambda v)^N\right) - \frac{v_{A3}^{N-M+1}}{N-M+1} g\left(\frac{N-M+1}{N}, -(\lambda v_{A3})^N\right) - \frac{v^{N+1}}{N+1} g\left(\frac{N+1}{N}, -(\lambda v)^N\right) + \frac{v_{A3}^{N+1}}{N+1} g\left(\frac{N+1}{N}, -(\lambda v_{A3})^N\right). \tag{5.14}$$

The above 12 equations, (5.3–5.14), are the h_c-based dimensionless GVF profiles expressed using the GHF with specified boundary conditions. Note that in these formulas $\lambda = h_c/h_n$, h_n is not real for flow in adverse channels. It is presumed to be of a "fictitious" value obtainable from the very same flow discharge, Q, in a "fictitious" channel on an identically sustaining (but negative) slope. $(x_{\sharp H2}, v_{H2})$, $(x_{\sharp H3}, v_{H3})$, $(x_{\sharp M1}, v_{M1})$, $(x_{\sharp M2}, v_{M2})$, $(x_{\sharp M3}, v_{M3})$, $(x_{\sharp C1}, v_{C1})$, $(x_{\sharp C3}, v_{C3})$, $(x_{\sharp S1}, v_{S1})$, $(x_{\sharp S2}, v_{S2})$, $(x_{\sharp S3}, v_{S3})$, $(x_{\sharp A2}, v_{A2})$ and $(x_{\sharp A3}, v_{A3})$ are specified boundary conditions for H2, H3, M1, M2, M3, C1, C3, S1, S2, S3, A2 and A3 profiles, respectively. The boundary conditions specified in plotting GVF profiles on all slopes, as shown in Figs. 5.3, 5.4, 5.5, 5.6, 5.7, are given in Table 5.2.

For illustration, we substitute the five N-values so assigned, one at a time, into each of the twelve particular solutions, (5.3–5.14) thereby plotting five h_c-based dimensionless GVF profiles on an H slope ($h_c/h_n = 0$) in Fig. 5.3, on an M slope ($h_c/h_n = 0.6$) in Fig. 5.4, on a C slope ($h_c/h_n = 1$) in Fig. 5.5, on an S slope ($h_c/h_n = 1.8$) in Fig. 5.6, and on an A slope ($h_c/h_n = 1/3$, in which h_n is fictitious) in Fig. 5.7. For comparison, the boundary conditions of the M and S profiles prescribed, respectively, in Figs. 5.4 and 5.6 for various N-values are, respectively, determined from those used in Figs 3.2 and 3.4 using the scaling relations, (4.7) and (4.8), for $N = 3$. However, the boundary conditions of the C profiles prescribed in Fig. 3.5 for various N-values, if determined from (4.7) and (4.8), are identical to those used in Fig. 3.2. As a result, the h_c-based C profiles so plotted are found to be exactly the same as the h_n-based C profiles because $h_c/h_n = 1$.

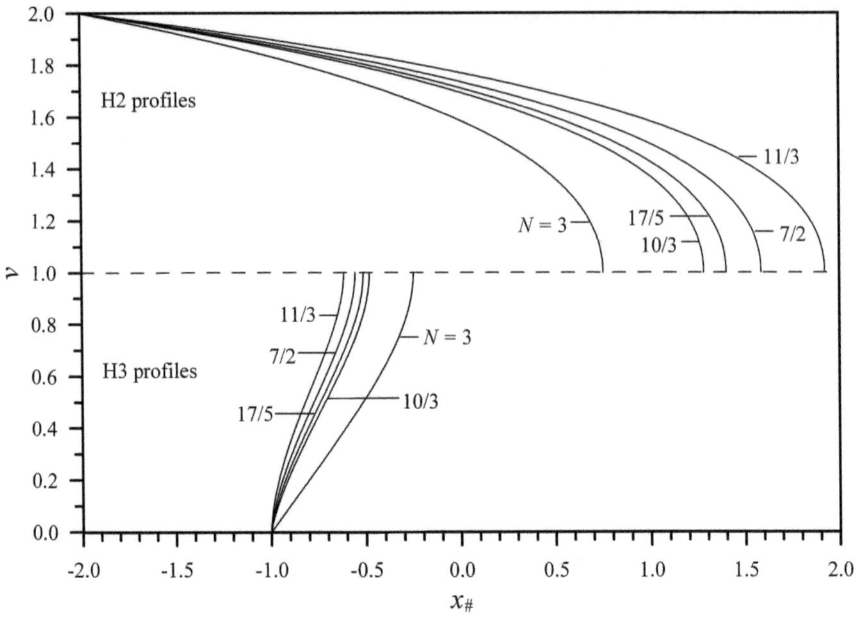

Fig. 5.3 The h_c-based dimensionless GVF profiles on a horizontal slope in a wide channel ($M = 3$). Two H profiles (H2 and H3) are plotted against various N-values

Comparison of the five h_c-based GVF profiles so plotted in each of the three zones on each type of the five channel slopes with one another may shed light on the effects of h_c/h_n (or its reciprocal, h_n/h_c) and the N-value on the features and properties of GVF profiles, such as their curvatures and inflection points. We will further address this and other related issues when such GHF-based solutions are analyzed later.

5.4 Conversion from the h_c-Based Solutions to the h_n-Based Solutions, or Vice Versa

5.4.1 *Solutions of GVF in Sustaining Channels*

Using (4.7) and (4.8), we can readily prove that the h_c-based solutions of (4.4), i.e., (4.32) [or (4.89) if $h_c/h_n = 1$] for $0 \le (h_c/h_n)v < 1$ (or $0 \le v < h_n/h_c$) and (4.47) [or (4.90) if $h_c/h_n = 1$] for $(h_c/h_n)v > 1$ (or $v > h_n/h_c$), can be converted to the h_n-based solutions of (3.5). In fact, to show the conversion from the former to the latter, Eqs. (4.32) and (4.47) upon substitution of $x_\sharp = (h_n/h_c)^{N+1}x_*$ [i.e., $x_* = \lambda^{N+1}x_\sharp$] from (4.8) and $v = (h_n/h_c)u$ [i.e., $u = \lambda v$] from (4.7) yield, respectively, the converted h_n-based solutions of (3.5) as follows:

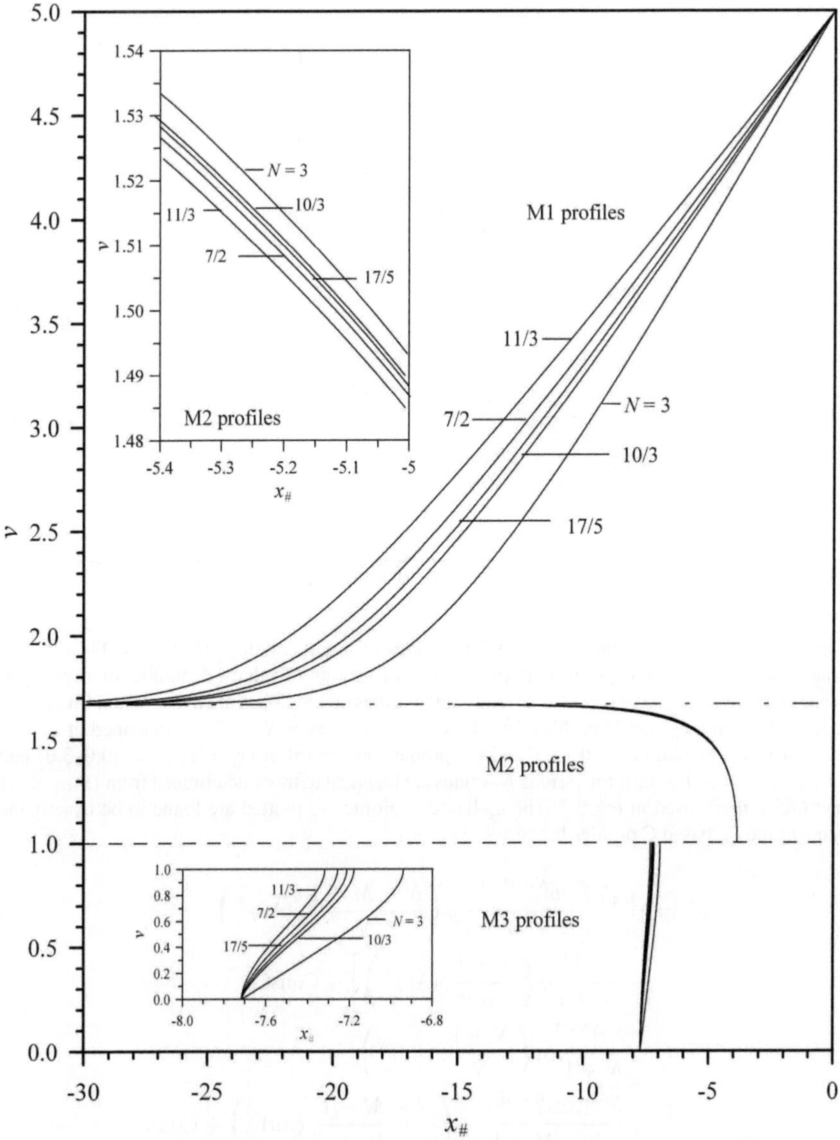

Fig. 5.4 The h_c-based dimensionless GVF profiles on a mild slope ($h_c/h_n = 0.6$) in a wide channel ($M = 3$). Three M profiles are plotted against various N-values. For comparison, the boundary conditions of the M1, M2 and M3 profiles prescribed at $(x_{\sharp M1}, v_{M1}) = (0.0, 5.0)$, $(x_{\sharp M2}, v_{M2}) = (-3.8580, 1.0)$ and $(x_{\sharp M3}, v_{M3}) = (-7.7160, 0.0)$ for various N-values are respectively determined from $(x_{*M1}, u_{M1}) = (0.0, 3.0)$, $(x_{*M2}, u_{M2}) = (-0.5, 0.6)$ and $(x_{*M3}, u_{M3}) = (-1.0, 0.0)$, used in Fig. 3.2, by means of Eqs. (4.7) and (4.8) for $N = 3$

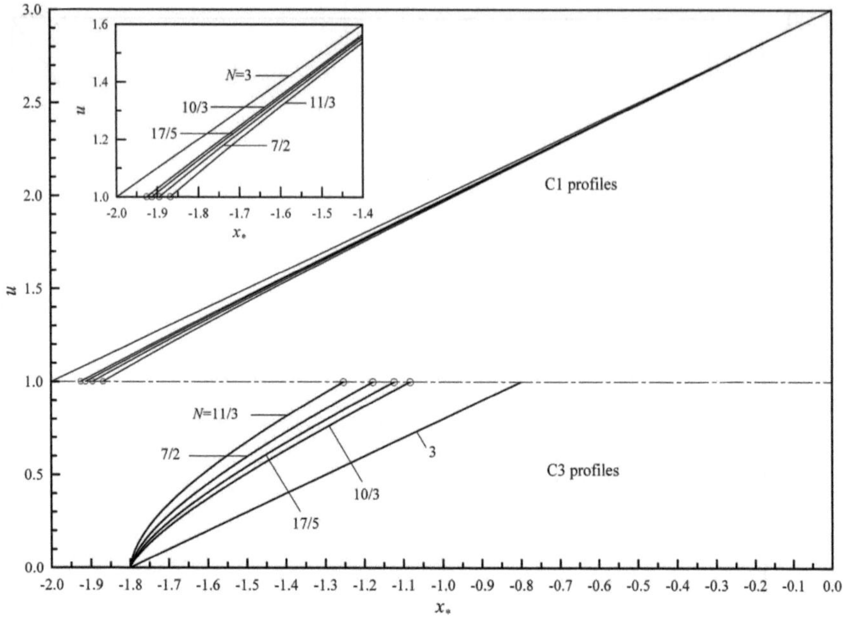

Fig. 5.5 The h_c-based dimensionless GVF profiles on a critical slope ($h_c/h_n = 1$) in a wide channel ($M = 3$). Two C profiles are plotted against various N-values. A number of *small open circles* plotted in the graph stand for so many singularities on the slopes of all the C1 and C3 profiles at $v = 1$, connoting that C profiles for all the N-values except $N = 3$ are undefined at $v = 1$. The boundary conditions of the C1 and C3 profiles prescribed at $(x_{\sharp C1}, v_{C1}) = (0.0, 3.0)$ and $(x_{\sharp C3}, v_{C3}) = (-1.8, 0.0)$ for various N-values are identical to those determined from (x_{*C1}, u_{C1}) and (x_{*C3}, u_{C3}), used in Fig. 3.3. The h_c-based C profiles so plotted are found to be exactly the same as the h_n-based C profiles because $h_c/h_n = 1$

$$
\begin{aligned}
x_* &= \lambda^{N+1}\left[\frac{v^{N-M+1}}{N-M+1} g\left(\frac{N-M+1}{N}, (\lambda v)^N\right)\right. \\
&\quad \left. - \frac{v^{N+1}}{N+1} g\left(\frac{N+1}{N}, (\lambda v)^N\right)\right] + \text{Const.} \\
&= -\frac{(\lambda v)^{N+1}}{N+1} g\left(\frac{N+1}{N}, (\lambda v)^N\right) \\
&\quad + \frac{\lambda^M (\lambda v)^{N-M+1}}{N-M+1} g\left(\frac{N-M+1}{N}, (\lambda v)^N\right) + \text{Const.} \\
&= \lambda v\left[1 - g\left(\frac{1}{N}, (\lambda v)^N\right)\right] \\
&\quad + \frac{\lambda^M (\lambda v)^{N-M+1}}{N-M+1} g\left(\frac{N-M+1}{N}, (\lambda v)^N\right) + \text{Const.} \\
&= u\left[1 - g\left(\frac{1}{N}, u^N\right)\right] \\
&\quad + \frac{\lambda^M u^{N-M+1}}{N-M+1} g\left(\frac{N-M+1}{N}, u^N\right) + \text{Const.},
\end{aligned}
\tag{5.15}
$$

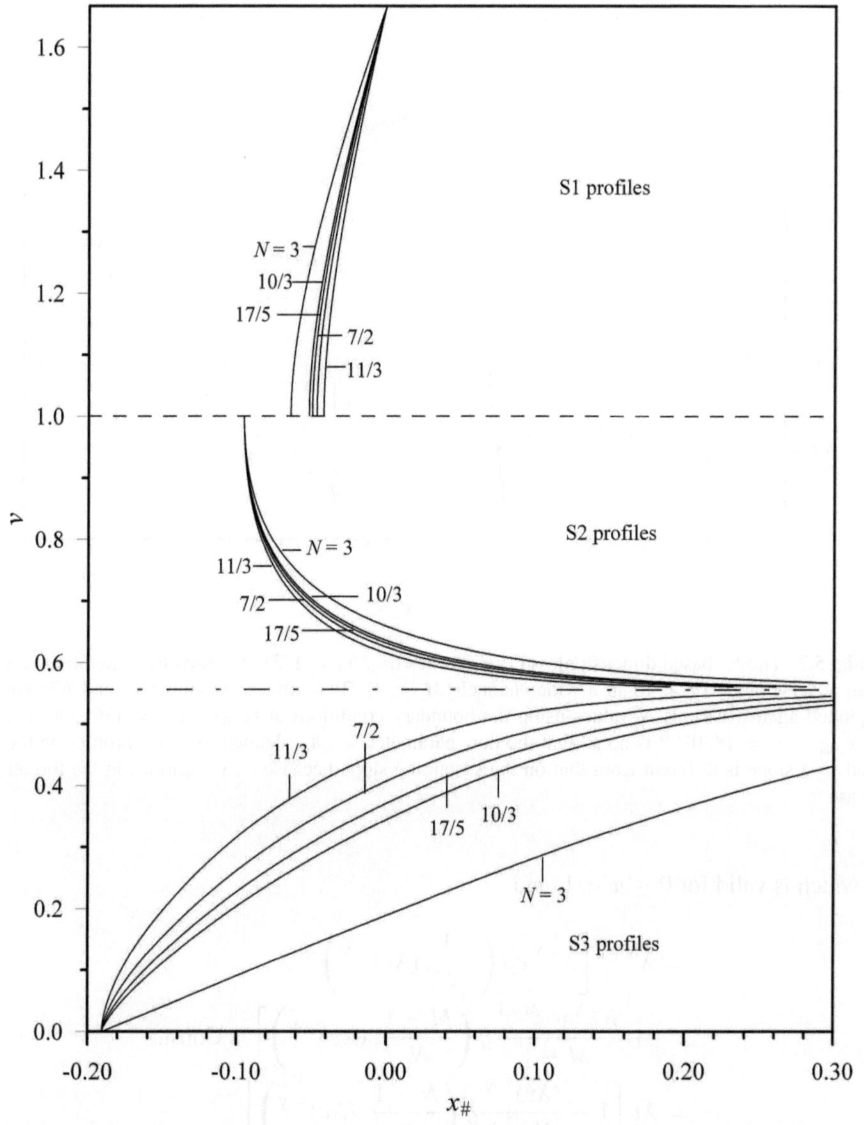

Fig. 5.6 The h_c-based dimensionless GVF profiles on a steep slope ($h_c/h_n = 1.8$) in a wide channel ($M = 3$). Three S profiles are plotted against various N-values. The boundary conditions for S1, S2 and S3 profiles prescribed at $(x_{\sharp S1}, v_{S1}) = (0.0, 1.6667)$, $(x_{\sharp S2}, v_{S2}) = (-0.0953, 1.0)$ and $(x_{\sharp S3}, v_{S3}) = (-0.1905, 0.0)$ are respectively determined from $(x_{*S1}, u_{S1}) = (0.0, 3.0)$, $(x_{*S2}, u_{S2}) = (-1.0, 1.8)$ and $(x_{*S3}, u_{S3}) = (-2.0, 0.0)$ used in Fig. 3.4, by means of Eqs. (4.7) and (4.8) for $N = 3$

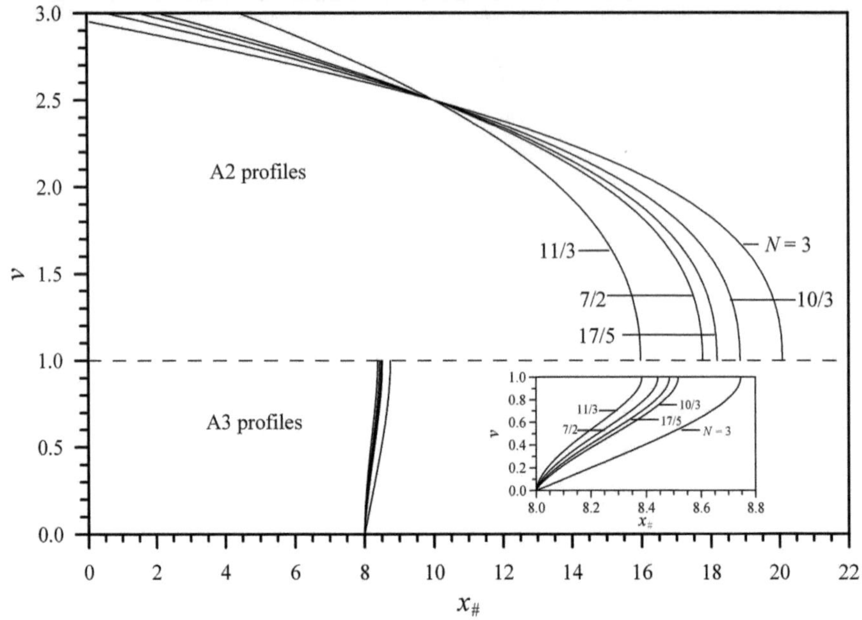

Fig. 5.7 The h_c-based dimensionless GVF profiles ($h_c/h_n = 1/3$), in which h_n is fictitious) on an adverse slope ($\theta < 0$) in a wide channel ($M = 3$). Two adverse profiles (A2 and A3) are plotted against various N-values using the boundary conditions at $(x_{\sharp A2}, v_{A2}) = (10, 2.5)$ and $(x_{\sharp A3}, v_{A3}) = (8, 0)$. It is noted that the flow parameter, h_c/h_n, defined for GVF profiles on the adverse slope is different from that on the sustaining slope because h_n is fictitious in the former case

which is valid for $0 \leq u < 1$, and

$$\begin{aligned} x_* &= \lambda^{N+1}\left[\lambda^{-N} v g\left(-\frac{1}{N}, (\lambda v)^{-N}\right)\right. \\ &\quad \left. + \frac{\lambda^{-N} v^{-M+1}}{M-1} g\left(\frac{M-1}{N}, (\lambda v)^{-N}\right)\right] + \text{Const.} \\ &= \lambda v\left[1 - \frac{(\lambda v)^{-N}}{N-1} g\left(\frac{N-1}{N}, (\lambda v)^{-N}\right)\right] \\ &\quad + \frac{\lambda^M (\lambda v)^{-M+1}}{M-1} g\left(\frac{M-1}{N}, (\lambda v)^{-N}\right) + \text{Const.} \\ &= u\left[1 - \frac{u^{-N}}{N-1} g\left(\frac{N-1}{N}, u^{-N}\right)\right] \\ &\quad + \frac{\lambda^M u^{-M+1}}{M-1} g\left(\frac{M-1}{N}, u^{-N}\right) + \text{Const.}, \end{aligned} \tag{5.16}$$

Table 5.2 Specified boundary conditions for h_c-based GVF profiles plotted in Figs. 5.3, 5.4, 5.5, 5.6, 5.7

Types of GVF profiles on a horizontal slope					
		H2		H3	
		$x_{\sharp H2}$	v_{H2}	$x_{\sharp H3}$	v_{H3}
		−2.0	2.0	−1.0	0.0
Types of GVF profiles on a mild slope					
M1		M2		M3	
$x_{\sharp M1}$	v_{M1}	$x_{\sharp M2}$	v_{M2}	$x_{\sharp M3}$	v_{M3}
0.0	5.0	−3.8580	1.0	−7.7160	0.0
Types of GVF profiles on a steep slope					
C1				C3	
$x_{\sharp C1}$	v_{C1}			$x_{\sharp C3}$	v_{C3}
0.0	3.0			−1.8	0.0
Types of GVF profiles on a steep slope					
S1		S2		S3	
$x_{\sharp S1}$	v_{S1}	$x_{\sharp S2}$	v_{S2}	$x_{\sharp S3}$	v_{S3}
0.0	1.6667	−0.0905	1.0	−0.1905	0.0
Types of GVF profiles on an adverse slope					
		A2		A3	
		$x_{\sharp H2}$	v_{H2}	$x_{\sharp H3}$	v_{H3}
		10.0	2.5	8.0	0.0

which is valid for $u > 1$. In the derivation processes of (5.15) and (5.16), we used the recurrence formulas between two contiguous GHF, as shown in (B.2) in Appendix B. The converted expressions of (4.32) and (4.47), i.e., (5.15) and (5.16), are exactly the h_n-based solutions of (3.5) that we obtain and are referred to as (3.30) and (3.43), respectively.

5.4.2 Solutions of GVF in Adverse Channels

Likewise, we can use (4.7) and (4.8) for flow in adverse channels to convert the h_c-based solutions of (4.12), i.e., (4.69) for $0 \leq (h_c/h_n)v < \infty$ (or $0 \leq v < \infty$), to the h_n-based solution of the counterpart of (4.12) based on h_n for $0 \leq u < \infty$, or vice versa. However, because the h_n-based solution of the counterpart of (4.12) based on h_n is beyond the scope of the Chap. 3, we first derive it as follows:

$$\frac{du}{dx_\sharp} = \frac{1 + u^N}{u^{N-M}(\lambda^M - u^M)}. \tag{5.17}$$

Equation (5.17) may be referred to as the h_n-based dimensionless GVF equation for flow in adverse channels with M, N and λ $(= h_c/h_n)$, as three parameters, in which h_n is fictitious. It is noted that (5.17) is distinguished from (3.2) by a sole difference in the "+" sign in front of the second term in the numerator on the right-hand side of (5.17). Instead of attempting to solve (5.17) for h_n-based dimensionless GVF profiles on the adverse slope, we simply use (4.7) and (4.8) to convert (4.69) to the h_n-based solutions of (5.17) as

$$\begin{aligned} x_* &= \frac{\lambda^{N+1} v^{N-M+1}}{N-M+1} g\left(\frac{N-M+1}{N}, -(\lambda v)^N\right) \\ &\quad - \frac{\lambda^{N+1} v^{N+1}}{N+1} g\left(\frac{N+1}{N}, -(\lambda v)^N\right) + \text{Const.} \\ &= \frac{\lambda^M u^{N-M+1}}{N-M+1} g\left(\frac{N-M+1}{N}, -u^N\right) \\ &\quad - \frac{u^{N+1}}{N+1} g\left(\frac{N+1}{N}, -u^N\right) + \text{Const.}, \end{aligned} \tag{5.18}$$

which should be valid for $0 \leq u < \infty$.

There is no intrinsic difference between the h_c-based solutions of (4.12) [i.e., (4.69)] and the h_n-based solution of (5.17) [i.e., (5.18)] because both solutions can be converted to or from each other using (4.7) and (4.8). Nevertheless, using (4.69) to compute the h_c-based GVF profiles on the A slope is more useful than (5.18) because (4.69) can reduce asymptotically to (4.11), i.e., the equation for computing the h_c-based dimensionless GVF profiles on the H slope, as λ ($= h_c/h_n$) $\rightarrow 0$ (see the proof in Appendix G), but (5.18) cannot asymptotically reduce to (4.11) as $\lambda \rightarrow 0$, i.e., not in the same way as (4.69) can, for the normalized variable u ($= h/h_n$) in (5.18) becomes undefined if $h_n \rightarrow \infty$.

For comparison, we plot (5.18) in Fig. 5.8 using the same $h_c/h_n = 1/3$ that is assumed in plotting (4.69) in Fig. 5.7. Besides the same h_c/h_n-value, we use the same boundary conditions specified in Table 5.2 for (4.69) to draw the h_n-based dimensionless GVF profiles on the A slope in Fig. 5.8 on the understanding that the boundary conditions for (5.18) are already converted from those for (4.69–4.7) and (4.8) for $N = 3$.

5.5 Discussion

It is expected that all features and properties of the h_c-based dimensionless GVF profiles in sustaining channels, such as the effects of both h_c/h_n and N-value on the slope of GVF profiles, its curvature and inflection points, and the existence of singularities, can reflect the very same of the corresponding h_n-based dimensionless GVF profiles in sustaining channels, as analyzed in Chap. 3, for the h_c-based dimensionless GVF profiles in sustaining channels, i.e., (4.32) and (4.47), can be converted to or from their respective h_n-based counterparts, i.e., (5.15) and (5.16).

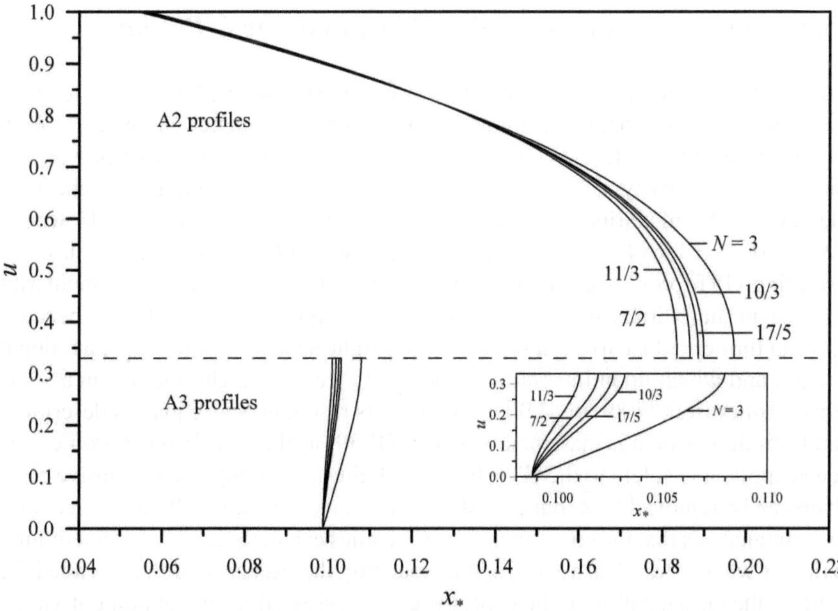

Fig. 5.8 The h_n-based dimensionless GVF profiles $h_c/h_n = 1/3$, in which h_n is fictitious) on an adverse slope ($\theta < 0$) in a wide channel ($M = 3$). Two adverse profiles (A2 and A3) are plotted against various N-values. The boundary conditions of A2 and A3 profiles prescribed at $(x_{*A2}, u_{A2}) = (0.1235, 0.8333)$ and $(x_{*A3}, u_{A3}) = (0.0988, 0)$ are determined from $(x_{\sharp A2}, v_{A2}) = (10, 2.5)$ and $(x_{\sharp A3}, v_{A3}) = (8, 0)$ used in Fig. 5.7 by means of Eqs. (4.7) and 4.8) for $N = 3$

Therefore, rather than iterating what have been reported in Chap. 3, we first analyze briefly below such features and properties from the standpoint of the h_c-based dimensionless GVF profiles in sustaining channels, and then focus on the scrutiny of all significant features and properties of the h_c-based dimensionless GVF profiles in adverse channels.

5.5.1 Features of the h_c-Based Dimensionless GVF Profiles

The right-hand side of (4.4), which represents the slope of the h_c-based dimensionless GVF profile in sustaining channels, is a rational function of v with N and $\lambda (= h_c/h_n)$ as two parameters. Evidently, $dv/dx_\sharp = \infty$ at $v = 1$ for all N-values; and so it is at $v = 0$ for all N-values except $N = M(= 3)$. By implication, a plot of $dv/dx_\sharp$ against v, as expressed in (4.4), possesses two vertical asymptotes at $v = 0$ and $v = 1$. In addition to the two vertical asymptotes, it has a horizontal asymptote at $v = \lambda^{-1}(= h_n/h_c)$ for various values of h_c/h_n and N, suggesting that the horizontal asymptote varies with the values of h_c/h_n and N except for $h_c/h_n = 1$ and $N = M(= 3)$.

5.5.2 Singularities of the h_c-Based Slopes of the C Profiles

It merits attention for $h_c/h_n = 1$ (i.e., for GVF on the C slope), (4.4) reduces to (5.2), which shows no horizontal asymptote because both denominator and numerator on the right-hand side of (4.4) become zero at $v = 1$ for all values of N except $N = M$, connoting a singularity at $v = 1$ in the rational function representing the right-hand side of (4.4). A singularity is displayed by an open small circle in Fig. 5.5. However, if $N = M = 3$ for (4.4) (i.e., a special case in which Chézy's formula is used in describing GVF profiles on the C slope in wide channels), there is no singularity at $v = 1$ in the rational function representing the right-hand side of (4.4) because (4.4) can further reduce to $dv/dx_\sharp = 1$, whose right-hand side is no longer a rational function and which upon integration yields (5.1), i.e., a straight line. It can also be derived either from (4.89) or (4.90). Although its resultant is an equation describing a uniform flow, which is classified by Chow (1959) as the so-called C2 profile, it is indeed applicable solely to the GVF for $N = 3$, thus pertaining to an exclusive case, which can be removed from our classification herein. In fact, it will be inferred later from our analysis that the curvature of the C profiles evaluated at $v = 1$ is flat infinite no matter what N-value is assumed, thus rejecting the existence of the so-called C2 profile if the classification of the C profile can be assessed from that point of view.

5.5.3 Two Inflection Points of the M Profiles

The inflection point of a flow profile on the $dv/dx_\sharp$ versus v plane takes place at a point, where $dv/dx_\sharp$ is an extremum (i.e., either a minimum or a maximum). It lies either between the two vertical asymptotes or between the horizontal asymptote and its adjacent vertical asymptote. The v-value, at which the inflection point occurs, can be determined from the condition under which $dv^2/dx_\sharp^2 = 0$. We can derive such a condition from (4.4) as:

$$dv^2/dx_\sharp^2 = -\frac{\left[1-\lambda^N v^N\right]\left[M\lambda^N v^N - N v^M + (N-M)\right]}{v^{2(N-M)+1}\left[1-v^M\right]^3}. \tag{5.19}$$

To cross-examine the derivation of (5.19), we substitute $x_\sharp = \lambda^{-(N+1)}x_*$ from (4.8) and $v = \lambda^{-1}u$ from (4.7) into (5.19), thereby converting the resultant to its counterpart on the du/dx_* versus u plane

$$du^2/dx_\sharp^2 = -\frac{(u^N-1)\left[M\lambda^N u^N - N u^M + (N-M)\lambda^M\right]}{u^{2(N-M)+1}\left[u^M-\lambda^M\right]^3}, \tag{5.20}$$

which is exactly what we derive in Chap. 3. Equating the right-hand side of the equal sign in (5.19) to zero yields the only possible condition under which $dv^2/dx_\sharp^2 = 0$, i.e.,

Table 5.3 The h_c-based dimensionless flow depths at two inflection points (IPs), one on the M1 profile and the other on the M3 profile, computed from (5.21) for the 4 N-values of GVF profiles on two mild slopes ($\lambda = 0.6$ and 0.95) in wide channels ($M = 3$)

Hydraulic exponents (M, N)	$\lambda = h_c/h_n = 0.6$		$\lambda = h_c/h_n = 0.95$		Given the values of $\lambda(< 1)$, M, and N to solve Eq. (5.21) or $M\lambda^N v^N - Nv^M + (N-M) = 0$
	v at IP on M1 profile $\lambda^{-1} < v < \infty$	v at IP on M3 profile $0 \le v \le 1$	v at IP on M1 profile $\lambda^{-1} < v < \infty$	v at IP on M3 profile $0 \le v \le 1$	
(3, 10/3)	226.861	0.4860	2.2296	0.6668	$3\lambda^{10/3}v^{10/3} - \frac{10}{3}v^3 + \frac{1}{3} = 0$
(3, 17/5)	105.101	0.5111	2.0424	0.6875	$3\lambda^{17/5}v^{17/5} - \frac{17}{5}v^3 + \frac{2}{5} = 0$
(3, 7/2)	48.622	0.5426	1.8640	0.7122	$3\lambda^{7/2}v^{7/2} - \frac{7}{2}v^3 + \frac{1}{2} = 0$
(3, 11/3)	22.433	0.5842	1.6916	0.7425	$3\lambda^{11/3}v^{11/3} - \frac{11}{3}v^3 + \frac{2}{3} = 0$

$$M\lambda^N v^N - Nv^M + (N - M) = 0. \tag{5.21}$$

For illustration, the two h_c-based solutions of v obtained from (5.21) for various N values under each of the assumed $h_c/h_n = 0.6$ and 0.95 are tabulated in Table 5.3, where (5.21) for $N = 10/3, 17/5, 7/2$, and $11/3$ are specifically listed in the table. It merits attention to remark in passing that Chen and Wang (1969) obtained exactly the very same expression of (5.21) for $N = 10/3$ (i.e., equivalent to using the Manning formula) although they used the normal-flow Froude number, $\mathbf{F_n}$ [$= (h_c/h_n)^{3/2}$], in place of h_c/h_n appearing in (5.21) to represent the flow characteristic at the normal-flow state for GVF in sustaining channels. In fact, (5.21) is the very same condition that can be converted from u to v through (4.7) using the following condition established in (5.20),

$$M\lambda^M u^N - Nu^M + (N - M)\lambda^M = 0. \tag{5.22}$$

In Chap. 3, we find from the four equations pertaining to (5.22), one for each of the four N-values studied under each of the assumed $h_c/h_n = 0.6$ and 0.95, the two solutions of u at the two inflection points of the h_n-based dimensionless GVF profiles on the M slope ($0 < h_c/h_n < 1$), one on the M1 profile ($u > 1$) and the other on the M3 profile ($0 < u \le h_c/h_n$) in wide ($M = 3$) channels. Correspondingly, we can solve each of the equations listed in Table 5.3 under each of the assumed $h_c/h_n = 0.6$ and 0.95 for the two values of v at the two inflection points of the h_c-based dimensionless GVF profiles on the M slope ($0 < h_c/h_n < 1$), one on the M1 profile ($v > h_n/h_c > 1$) and the other on the M3 profile ($0 < v \le 1 \le h_n/h_c$). It is noted that the two h_c-based solutions of v so obtained are identical to the corresponding two h_n-based solutions of u obtained from each of the four equations pertaining to (5.22) for all the N-values studied except $N = 3$ because such solutions expressed in terms of v can be converted to those in terms of u via (4.7) or vice versa. The solutions of v listed in Table 5.3 check exactly with their counterparts expressed in terms of u in Table 3.7 though the h_n-based condition for the existence of the inflection points, (5.22), breaks down as $\lambda = h_c/h_n \to 0$ because $u(= h/h_n)$ used in (5.22) becomes undefined as $h_n \to \infty$. On the contrary, (5.21) can asymptotically reduce to the sole solution of v at the inflection point of the H3 profile on the horizontal slope as $h_c/h_n \to 0$, as shown in the following subsection. Equation (5.21) is thus valid in the range of $0 \le h_c/h_n < 1$, including one end point at $h_c/h_n = 0$ but excluding the other end point at $h_c/h_n = 1$.

In Fig. 5.9, we plot eight solution curves to trace the paths of the two inflection points of the M1 and M3 profiles on the v versus h_c/h_n plane with N as a parameter for all the N-values studied except $N = 3$, thereby showing the combined effects of h_c/h_n and N on the locations of the two inflection points. It merits mentioning that one of the distinctive advantages of the h_c-based dimensionless GVF equation, (4.4), over the corresponding h_n-based dimensionless GVF equation, (3.5), is reflected in Fig. 5.9. This figure shows that (5.21) can asymptotically reduce to the sole solution of v at the inflection point of the H3 profile as $h_c/h_n \to 0$, but (5.22) cannot as a result of the breakdown in the limit of u $(= h/h_n)$ as $h_n \to \infty$. Brought up in the following for further discussion is the asymptotic reduction of (5.21) to the inflection point of the H3 profile as $h_c/h_n \to 0$.

5.5.4 *Sole Inflection Point of the H3 Profile*

Given the values of h_c/h_n and N, we can use either (5.20) or (5.21) to find the two inflection points of an h_c-based or h_n-based dimensionless M profile, as pointed out above. Nevertheless, we cannot formulate or plot the h_n-based dimensionless GVF profiles on a horizontal slope, let alone use (5.21) to locate the sole inflection point of an h_n-based dimensionless H3 profile. On the contrary, we can not only use (4.11) to plot the h_c-based dimensionless H profiles for all the N-values studied, as shown in Fig. 5.3, but also use (5.20) to determine the sole inflection point of an h_c-based dimensionless H3 profile. To locate the sole inflection point is straightforward; i.e., (5.21) on substitution of $h_c/h_n = 0$ yields

$$v = \left(\frac{N - M}{N}\right)^{1/M}, \tag{5.23}$$

which is valid for all the N-values studied except $N = 3$ in wide channels ($M = 3$). In Fig. 5.3, the H3 profiles are plotted for $N = 3$, 10/3, 17/5, 7/2, and 11/3, while the v-values at the inflection points ($v_{i.p.}$) for all such N-values except $N = 3$ can be computed from (5.23) to be $v_{i.p.} = \{0.46416, 0.49000, \ 0.52276, 0.56652\}$,

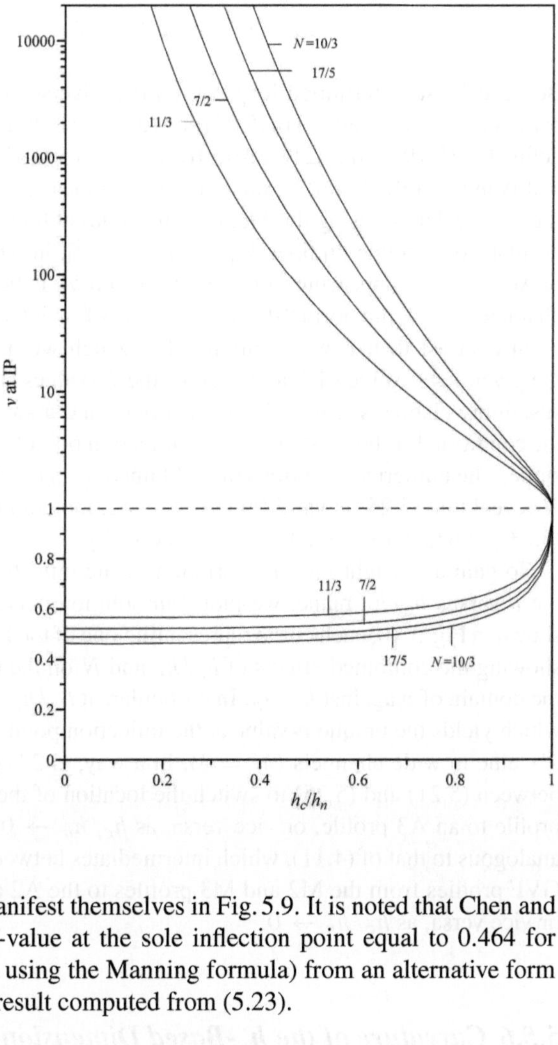

Fig. 5.9 Plot of eight solution curves representing the two values of v at the inflection points (IP), one on each of the M1 and M3 profiles, against the various values of h_c/h_n under $N = 10/3, 17/5, 7/2,$ and $11/3$. For clarity in displaying the path of the IP of the M1 profiles, which spread over an enormously wide range of v with respect to h_c/h_n, as compared with that of the M3 profiles, which crowd in a narrow range of v, we adopt two composite scales to mark ticks on the ordinate, namely the linear scaling for $0 \le v < 1$ and the logarithmic scaling for $v > 1$. Jan and Chen (2013), reproduced with permission

respectively, which should manifest themselves in Fig. 5.9. It is noted that Chen and Wang (1969) obtained the v-value at the sole inflection point equal to 0.464 for $N = 10/3$ (i.e., equivalent to using the Manning formula) from an alternative form of (4.9), thus confirming the result computed from (5.23).

5.5.5 Sole Inflection Point of the A3 Profile

Using (4.12), we can determine the v-value, at which the inflection point of a GVF profile on the adverse slope takes place, from the condition under which $d^2v/dx_\sharp^2 = 0$,

$$dv^2/dx_\sharp^2 = \frac{[1+\lambda^N v^N][M\lambda^N v^N + Nv^M - (N-M)]}{v^{2(N-M)+1}(1-v^M)^3}. \tag{5.24}$$

Equating the right-hand side of the equal sign in (5.24) to zero yields the only feasible condition under which we can have $d^2v/dx_\sharp^2 = 0$, i.e.,

$$M\lambda^N v^N + N v^M - (N - M) = 0. \tag{5.25}$$

Because h_n so determined for GVF on the adverse slope is fictitious and so is the parameter, h_c/h_n, defined in (5.24) and (5.25), we should not mix them up with those defined in (5.19) and (5.21), respectively. It is worthwhile to remark again that Chen and Wang (1969) obtained exactly the very same expression of (5.25) for $N = 10/3$ (i.e., equivalent to using the Manning formula) although they used the "conceptual" normal-flow Froude number, $\mathbf{F}_\mathbf{n}^*[= (h_c/h_n)^{3/2}]$, in which h_n is fictitious], in place of $\lambda(= h_c/h_n)$ appearing in (5.25) (in which h_n is fictitious) to represent the flow characteristic at the normal-flow state for GVF in adverse channels.

In contrast to the two solutions of v which we have obtained from (5.21) for the given valid value of λ and each of the N-values studied except $N = 3$ in wide sustaining channels ($M = 3$), we can only find a sole solution of v in (5.25), i.e., the condition for the existence of the inflection point for GVF profiles on the adverse slope. The numerical solutions of v obtained from (5.25) for various values of λ and N reveal that (5.25) is valid for $0 \leq \lambda < \infty$, including one end point at $\lambda = 0$, i.e., (5.23), to which (5.25) reduces asymptotically.

To gain an insight into the variation of the inflection point of the A3 profile on the v versus h_c/h_n plane, we plot four solution curves with N as a parameter, as shown in Fig. 5.10, each curve representing one of the four N-values studied, thereby showing the combined effects of h_c/h_n and N on the path of the inflection point in the domain of v against h_c/h_n. In particular, at $h_c/h_n = 0$, (5.25) reduces to (5.23), which yields the unique v-value at the inflection point of an H3 profile for the given N-value in wide channels ($M = 3$). In a way, (5.23) plays the role of an interface between (5.21) and (5.25) to switch the location of the inflection point from an M3 profile to an A3 profile, or vice versa, as $h_c/h_n \to 0$. In fact, the role of (5.21) is analogous to that of (4.11), which intermediates between (4.32) and (4.69) to switch GVF profiles from the M2 and M3 profiles to the A2 and A3 profiles, respectively, or vice versa, as $h_c/h_n \to 0$.

5.5.6 Curvature of the h_c-Based Dimensionless GVF Profiles

The curvature, K_v, of the h_c-based dimensionless GVF profiles at any point ($x_\sharp$, v) can be expressed from calculus as

$$K_v = \frac{\left|\dfrac{d^2 v}{dx_\sharp^2}\right|}{\left[1 + \left(\dfrac{dv}{dx_\sharp}\right)^2\right]^{3/2}}. \tag{5.26}$$

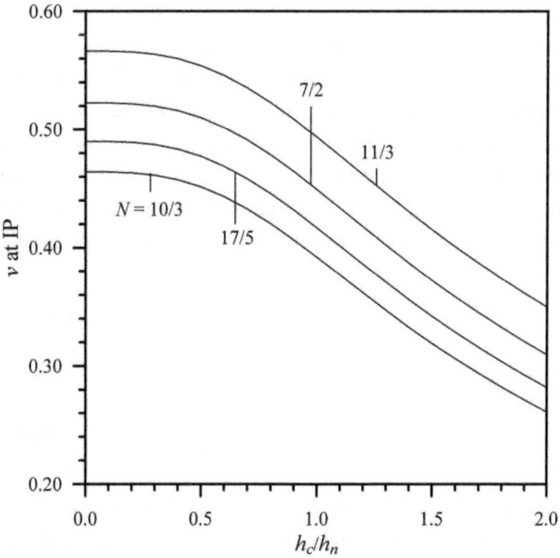

Fig. 5.10 Plot of the curvature, K_v, at $v = 1$ against h_c/h_n with the N-value as a parameter for GVF profiles in sustaining channels and also in adverse channels. The theoretical range of h_c/h_n in this plot is $0 \leq h_c/h_n < \infty$, but it is only plotted for $0 \leq h_c/h_n < 2$ herein. Jan and Chen (2013), reproduced with permission

The K_v of the h_c-based dimensionless one-dimensional GVF profiles in adverse channels should be different from that in sustaining channels because of the difference in the governing equations used to describe the respective GVF profiles in both channels. The K_v for the h_c-based dimensionless one-dimensional GVF profiles in sustaining channels is expressed by (5.26) on substitution of $d^2v/dx_\sharp^2$ from (5.19) and $dv/dx_\sharp$ from (4.4) as

$$K_v = \frac{\mid [1 - \lambda^N v^N][M\lambda^N v^N - Nv^M + (N - M)] \mid}{\mid v^{-N+M+1} \mid \left[[v^{N-M}(1 - v^M)]^2 + [1 - \lambda^N v^N]^2 \right]^{3/2}} \tag{5.27}$$

and K_v for the h_c-based dimensionless GVF profiles in adverse channels is expressed by (5.26) on substitution of $d^2v/dx_\sharp^2$ from (5.24) and $dv/dx_\sharp$ from (4.12) as

$$K_v = \frac{\mid [1 + \lambda^N v^N][M\lambda^N v^N + Nv^M - (N - M)] \mid}{\mid v^{-N+M+1} \mid \left[[v^{N-M}(1 - v^M)]^2 + [1 + \lambda^N v^N]^2 \right]^{3/2}}. \tag{5.28}$$

Evidently, (5.27) and (5.28) show that K_v is zero at the inflection point by virtue of (5.21) and (5.25), respectively; and so is that K_v is zero at the place where the GVF profile becomes parallel with the bed at $h = h_n$ (i.e., $v = h_n/h_c$) in sustaining channels as a result of the zero factor in the numerator of (5.27), but not in adverse channels owing to the nonzero factor in the numerator of (5.28). The latter consequences may be also self-evident from $dv/dx_\sharp = 0$ in (4.4) at $v = h_n/h_c$, but $dv/dx_\sharp \neq 0$ in (4.12) at $v = h_n/h_c$. Because $N - M > 0$ and $-N + M + 1 > 0$ for all the N-values studied, it can be readily inferred that $dv/dx_\sharp = \infty$ and $K_v = \infty$ for all profiles at the channel bed (i.e., $v = 0$) in sustaining channels by virtue of (4.4) and (5.27) as well as in adverse channels by virtue of (4.12) and (5.28). On the other hand, (4.4) and (4.12) show that $dv/dx_\sharp = \infty$ at $v = 1$, though the values of K_v expressed by (5.27) and (5.28) should be relatively large, but not infinity, for every profile, as illustrated in Fig. 5.1 for $N = 10/3$. It is thus worthwhile to compare K_v at such a specific point on the GVF profile. For example, for K_v at $v = 1$, (5.27) and (5.28), upon substitution of $v = 1$, yields, respectively,

$$K_v \mid_{v=1} = \frac{M}{\mid \lambda^N - 1 \mid} \quad \text{(in sustaining channels)} \tag{5.29}$$

and

$$K_v \mid_{v=1} = \frac{M}{\mid \lambda^N + 1 \mid} \quad \text{(in adverse channels).} \tag{5.30}$$

Equations (5.29) and (5.30) are the expressions of K_v at $v = 1$ for GVF profiles in sustaining channels and in adverse channels, respectively, indicating that for the given N-value it further depends on the value of λ. Obviously, if λ is zero, namely for the H profiles, we obtain K_v at $v = 1$ equal to M $(= 3)$ both from (5.29) and (5.30). On the other hand, if λ $(= h_c/h_n)$ increases from zero and approaches unity, K_v at $v = 1$ increases correspondingly and approaches infinity as a limit from (5.29) for GVF profiles in sustaining channels, whereas K_v at $v = 1$ for the A profiles decreases gradually from $M(= 3)$ and becomes $M/2$ $(= 1.5)$ from (5.30) as $h_c/h_n = 1$. For gaining an insight into the effects of h_c/h_n and N on K_v at $v = 1$ at a glance, we use (5.29) and (5.30) to plot K_v at $v = 1$ against h_c/h_n with the N-value as a parameter, as shown in Fig. 5.11.

Inspection of Fig. 5.11 reveals that K_v at $v = 1$ for GVF profiles in sustaining channels approaches infinity as $h_c/h_n \to 1$. This trend for K_v at $v = 1$ implies that the closer the two v-values at the two inflection points of the M profiles approach unity, as shown in Fig. 5.9, the larger is K_v at $v = 1$, irrespective of the M or S profile, as shown in Fig. 5.11. In contrast, K_v at $v = 1$ for A profiles in adverse channels decreases and approaches zero as $h_c/h_n \to \infty$ [as per (5.30)]. The A profile corresponding to this limit $(h_c/h_n = \infty)$ is a vertical line with its K_v everywhere along the line is zero, as manifesting itself in Fig. 5.1. As for the effect of N on K_v at $v = 1$, Fig. 5.11 also shows that the smaller the N-value, the larger is K_v at $v = 1$ for all M, S, and A profiles in the whole range of h_c/h_n, except that it tends to behave opposite for A profiles spanning $0 \le h_c/h_n \le 1$.

To have a good grip on the effect of h_c/h_n on K_v at $v = 1$, attention is refocused on one of the most critical limits for M and S profiles in sustaining channels as $h_c/h_n \to 1$, where K_v at $v = 1$ is infinite [as per (5.29)] and the C profiles are undefined at $v = 1$ because both M and S profiles collapse onto the C1 and C3 profiles by creating the two singularities $v = 1$ as $h_c/h_n \to 1$. It is thus meaningless to evaluate K_v of the C profiles at $v = 1$; and so it is by implication to classify the C profiles at $v = 1$ as the so-called C2 profiles by Chow (1959), no matter what N-value in (5.29) is given as long as $h_c/h_n = 1$.

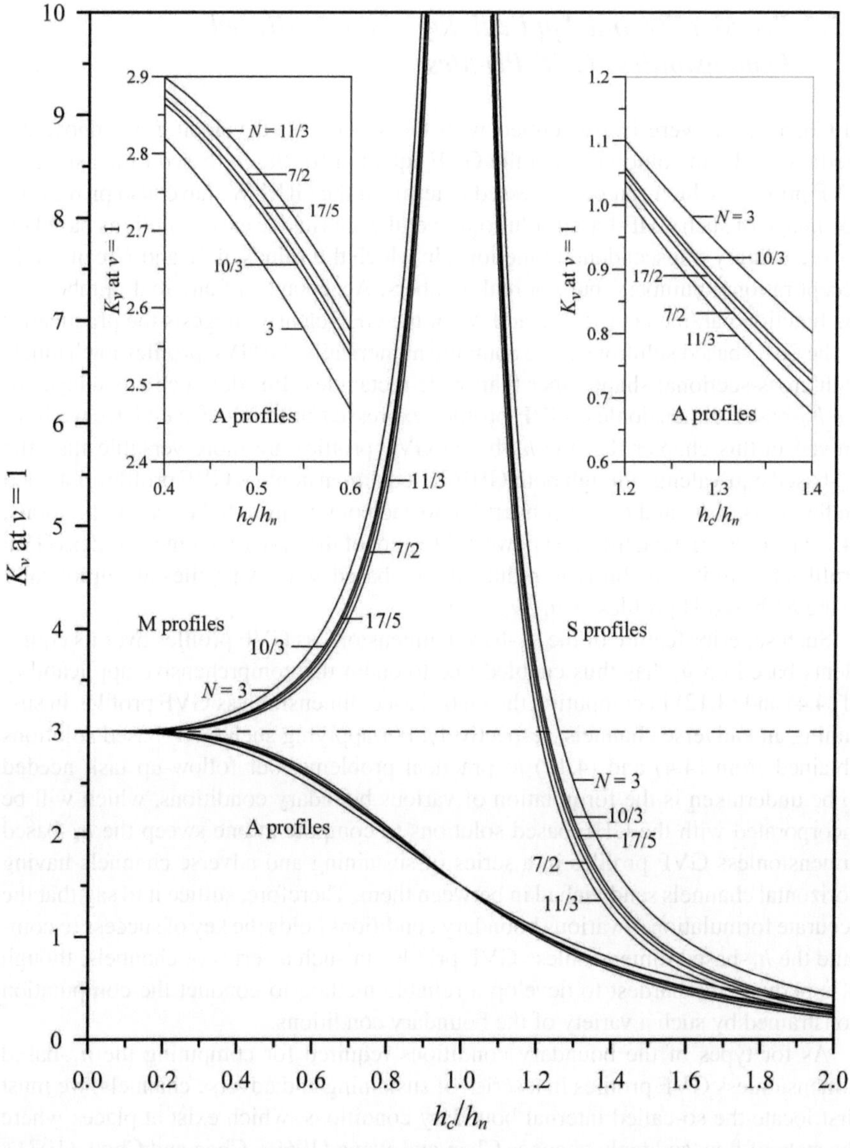

Fig. 5.11 Plot of four solution curves, each representing the single value of v at the inflection point (IP) of the A3 profile against the various values of h_c/h_n (in which h_n is fictitious) under $N = 10/3, 17/5, 7/2$, and $11/3$. The theoretical range of h_c/h_n is $0 \leq h_c/h_n < \infty$, but for simplicity it is only plotted for $0 \leq h_c/h_n < 2$ herein. Jan and Chen (2013), reproduced with permission

5.5.7 Practicality and Applicability of the h_c-Based Dimensionless GVF Profiles

In Chap. 3, we were overwhelmed with the success in developing an innovative method to obtain solutions from the GVF equation for the h_n-based dimensionless GVF profiles, which can be expressed in terms of the GHF. We have also proved the advantage of such GHF-based solutions over the alternative exact solutions based on the elementary transcendental functions, in which the values of M and N can solely accept rational numbers, but not real numbers. Acceptance of any real numbers by the function parameters (i.e., M and N) of the GHF clearly suggests the practicality of the GHF-based solutions if we compute numerically the GVF profiles in channels with cross-sectional shape other than wide rectangles. Besides such advantages of the h_n-based dimensionless GVF profiles expressed in terms of the GHF, we have proved in this chapter that the h_c-based GVF profiles are more versatile than the h_n-based equivalents, though both GHF-based dimensionless GVF profiles obtained on the basis of h_n and h_c are convertible to each other through the scaling relations, (4.7) and (4.8). In fact, the most powerful feature of the h_c-based dimensionless GVF profiles lies in its capability to reduce the h_c-based M or A profiles asymptotically to the h_c-based H profiles as $h_c/h_n \to 0$.

Such superior feature of the h_c-based dimensionless GVF profiles over its equivalents based on h_n has thus enabled one to claim the comprehensive applicability of (4.4) and (4.12) in computing the GHF-based dimensionless GVF profiles in sustaining and adverse channels, respectively. For applying such GHF-based solutions obtained from (4.4) and (4.12) to practical problems, our follow-up task needed to be undertaken is the formulation of various boundary conditions, which will be incorporated with the GHF-based solutions to compute at one sweep the h_c-based dimensionless GVF profiles in a series of sustaining and adverse channels having horizontal channels sandwiched in between them. Therefore, suffice it to say that the accurate formulation of various boundary conditions holds the key of success to compute the h_c-based dimensionless GVF profiles in such a series of channels, though it is perhaps the hardest to develop a reliable method to conduct the computation constrained by such a variety of the boundary conditions.

As for types of the boundary conditions required for computing the h_c-based dimensionless GVF profiles in a series of sustaining and adverse channels, we must first locate the so-called internal boundary conditions, which exist at places where the state of flow suddenly changes Chen and Wang (1969), Chen and Chow (1971). One type of the internal boundary conditions needed is at hydraulic jumps and overfalls, which occur in prismatic channels at places where the flow condition changes rapidly from a supercritical state to a subcritical state and vice versa under "freely" flowing conditions. The other type of internal boundary conditions needed is at sudden or rapid transitions in channel width and cross-sectional shape under "forced" flowing conditions as a result of constricted flows passing through hydraulic structures, such as weirs and sluice gates built in non-prismatic channels. Apparently, there exist various types of internal boundary conditions, such as the hydraulic-jump equations at places where hydraulic jumps occur, the formation of the critical Froude number at places where overfalls are induced, and the calibrated discharge relations (or rating curves) at places where weirs and sluice gates among other discharge-measuring devices are installed. In fact, it is quite involved to compute the h_c-based dimensionless GVF profiles subject to such a variety of the internal boundary conditions imposed at many places as needed in a series of artificially or naturally formed prismatic and non-prismatic channels.

5.6 Summary

This chapter has laid the foundation to compute at one sweep the h_c-based dimensionless GVF profiles in a series of sustaining and adverse channels, which have horizontal slopes sandwiched in between them. To obtain the GHF-based dimensionless solutions from the h_c-based GVF equation is our first step for developing a viable method to compute the h_c-based dimensionless GVF profiles subject to a variety of the boundary conditions imposed in such a series of interconnected sustaining and adverse channels. Working toward that goal, we have come up with two significant results produced from this study: Firstly, we have obtained the GHF-based solutions from the h_c-based dimensionless GVF equation, which proves to be applicable for computing the GVF profiles in both sustaining and adverse channels. Secondly, we have analytically proved that the GHF-based dimensionless M and A profiles, if normalized by h_c rather than by h_n, can asymptotically reduce to the h_c-based dimensionless H profiles as $h_c/h_n \to 0$. Both significant results thus constitute the principal conclusions drawn from this study.

In practical applications, the h_c-based dimensionless GVF profiles expressed in terms of the GHF can prove to be more useful and versatile than the h_n-based equivalents obtained in Chap. 3 though both profiles are convertible to each other through the scaling relations, (4.7) and (4.8). Among the well-known advantages of such h_c-based dimensionless GVF profiles over their counterparts based on h_n, there lies the most powerful feature of the h_c-based GVF profiles expressed in terms of the GHF, with which one can readily reduce the h_c-based M and A profiles asymptotically to the h_c-based H profiles as $h_c/h_n \to 0$. In fact, we have proved that the M2 and M3 profiles can asymptotically reduce to the H2 and H3 profiles, respectively, as $h_c/h_n \to 0$; and so can the A2 and A3 profiles to the H2 and H3 profiles, respectively. However, contrary to the h_c-based M and A profiles, the h_c-based C or S profiles cannot be proved to reduce asymptotically to the h_c-based H profiles as $h_c/h_n \to 0$. In a way, such asymptotic nature of the h_c-based dimensionless GVF equation and the solutions thereof has enabled one to remove the traditional, yet redundant assumption of small θ imposed in the derivation of the GVF equation.

After decades of long struggle by hydraulicians in their attempts to improve the rudimentary approach taken to solve the GVF equation using h_n and the so-called varied-flow function, we have finally come up with a novel approach to solve the same problem based on h_c and the GHF instead. As shown in Fig. 5.1, an innovated formulation of the h_c-based dimensionless GVF profiles expressed in terms of the GHF has greatly advanced the conventional technique used in the GVF computation to the extent that hydraulicians for the first time in the computer age can fully utilize a mathematics software, which is capable of producing the GHF-based solutions of the h_c-based dimensionless GVF equation. The principal conclusions so drawn from this study embrace all significant results acquired from the in-depth analysis of the h_c-based dimensionless solutions expressed in terms of the GHF along with those attained in the exact proof of the asymptotic reduction of the h_c-based dimensionless M and A profiles to the corresponding H profiles as $h_c/h_n \to 0$.

Included among the other conclusions drawn from this study are some findings and minor results, which are recapped in order of presentation as follows:

1. A plot of the GHF-based solutions on the $(x_\sharp, v)$-plane with h_c/h_n as a parameter helps gaining an insight into the uniqueness and versatility of such GHF-based solutions. On prescription of the boundary conditions, one for each solution, we can determine the constants of integration, "Const.", in (4.32), (4.47), and (4.69), thereby plotting these equations on the $(x_\sharp, v)$-plane with h_c/h_n as a parameter. For clarity, we have plotted Fig. 5.1, in which the h_c-based dimensionless GVF profiles in sustaining and adverse channels for $M = 3$ and $N = 10/3$ are shown against the various values of h_c/h_n.
2. Inspection of Fig. 5.1 reveals that the particular solution curves drawn by (4.32) and (4.47) with $h_c/h_n = 1$ separate the region of M profiles from that of S profiles, while the unique solution curve drawn by (4.11) through the asymptotic reduction of (4.32) or (4.69) as $h_c/h_n \to 0$ separates the region of M profiles from that of A profiles. These two solution curves are respectively referred to as the critical (C) profiles on the critical (C) slope ($h_c/h_n = 1$) and the horizontal (H) profiles on the horizontal (H) slope ($h_c/h_n = 0$).
3. The particular solutions of (4.32) and (4.47) on substitution of $h_c/h_n = 1$ yield (4.89) and (4.90), which are, respectively, referred to as the equations for h_c-based C3 profiles in zone 3 ($0 \le v < 1$) and C1 profiles in zone 1 ($1 < v < \infty$). In particular, (4.89) upon substitution of $M = N = 3$ readily reduces to (5.1) with the help of a recurrence formula (B.4), which is a straight line. Likewise, (4.90) on substitution of $M = N = 3$ also reduces to (5.1), a straight line, but it differs in the expressed constant of integration from (4.89) due to the difference in the specified boundary conditions. It is found that (4.89) and (4.90) are all undefined at $v = 1$, connoting the existence of a singularity at $v = 1$. However, if $M = N = 3$, there is no singularity at $v = 1$.
4. As shown in Fig. 5.1, a solution curve plotted by use of (4.11), to which (4.32) and (4.69) reduce asymptotically as $h_c/h_n \to 0$ from either side of (4.11) separates the region of M profiles from that of A profiles. We have proved in Appendix G that (4.11) can be obtained by the asymptotic reduction of (4.32) or (4.69) as $h_c/h_n \to 0$, using the GHF definition. In fact, not all the solution curves described by (4.32) can asymptotically reduce to the H2 profile in zone 2 ($1 \le v < \infty$) or the H3 profile in zone 3 ($0 \le v \le 1$) as $h_c/h_n \to 0$ because the limits of the first and second terms on the right-hand side of (4.32) at $h_c/h_n = 0$

for some profiles, such as the C3 profile ($0 \le v < 1$) plotted by (4.89) and the S3 profiles ($0 \le v < h_n/h_c$) on the S slope ($1 < h_c/h_n < \infty$), do not exist. These results are summarized in Table 5.1

5. The horizontal asymptotes at $v = h_n/h_c$ on the $(x_\sharp, v)$-plane for the given h_n/h_c-values in the domain of solution curves plotted for (4.32) and (4.47) are deemed as "movable" lines of demarcation drawn to divide the domain of v into two major regions: One region spans $0 \le (h_c/h_n)v < 1$ (or $0 \le v < h_n/h_c$), which covers both zones 2 and 3 for M profiles as well as zone 3 for S profiles, and the other region extends over $(h_c/h_n)v > 1$ (or $v > h_n/h_c$), which covers zone 1 for M profiles as well as both zones 1 and 2 for S profiles. In contrast, there exist no horizontal asymptotes at $v = h_n/h_c$ for various h_n/h_c-values in the domain of solution curves plotted by (4.69) on the $(x_\sharp, v)$-plane, representing the h_c-based dimensionless GVF profiles on the adverse slope because h_n adopted in (4.69) is simply fictitious (i.e., nonexistent).
6. In contrast to eight types of the h_n-based GVF profiles, as classified in Chap. 3, there exist a total of twelve types of the h_c-based GVF profiles, two on the H slope ($h_c/h_n = 0$), three on the M slope ($0 < h_c/h_n < 1$), two on the C slope ($h_c/h_n = 1$), three on the S slope ($1 < h_c/h_n < \infty$), and two on the A slope ($0 \le h_c/h_n < \infty$, in which h_n is fictitious). The twelve h_c-based GVF profiles in three zones may be, respectively, referred to as H2 and H3; M1, M2, and M3; C1 and C3; S1, S2, and S3; A2 and A3. The twelve profiles so classified are the same as those classified by Chow (1959) except for C2, which may be excluded from our classification because it is a singularity for all the N-values except for $N = 3$.
7. Using (4.69) to compute the h_c-based dimensionless GVF profiles on the A slope is more useful than (5.18), i.e., the equation for computing the h_n-based counterpart on the A slope, because (4.69) can reduce asymptotically to (4.11), i.e., the equation for computing the h_c-based dimensionless GVF profiles on the H slope, as $h_c/h_n \to 0$ (see the proof in Appendix G), but (5.18) cannot asymptotically reduce to (4.11) as $h_c/h_n \to 0$, at least not in the same way as (4.69), for the normalized variable $u(= h/h_n)$ in (5.18) becomes undefined if $h_n \to \infty$.
8. In case of $h_c/h_n = 1$, (4.4) shows no horizontal asymptote because both denominator and numerator on the right-hand side of (4.4) become zero at $v = 1$ for all values of N except $N = 3$, connoting a singularity at $v = 1$ in the rational function representing the right-hand side of (4.4). The classification of the C profile can be assessed from the viewpoint of the curvature on the C profiles evaluated at $v = 1$ because it is simply infinite no matter what N-value is assumed, thus rejecting the existence of the so-called C2 profile.
9. We have formulated (5.21), (5.23), and (5.25), i.e., the conditions for locating the values of v at the inflection points of the h_c-based dimensionless GVF profiles on the M, H, and A slopes, respectively. In other words, given a set of the values of h_c/h_n and N, we can solve such equations, respectively, for two inflection points on the M profiles (i.e., one on each of the M1 and M3 profiles) and a sole inflection point on each of the H3 and A3 profiles. Plotted in Figs. 5.9 and

5.10 are solution curves for the respective inflection points on such profiles. In a way, using both figures, we can show that (5.23) plays a role of the intermediate between (5.21) and (5.25) to switch the location of the inflection point from the M3 profile to the corresponding A3 profile, or vice versa, as $h_c/h_n \rightarrow 0$.

10. It is found that the curvature, K_v, of the h_c-based dimensionless GVF profiles in adverse channels differs a little bit from that in sustaining channels because the governing equations used in the description of the respective GVF profiles in both channels are slightly different. For example, K_v is zero at the inflection point; and so is it at the place where the GVF profile becomes parallel with the bed at $h = h_n$ (or $v = h_n/h_c$) in sustaining channels, but not in adverse channels. It can be readily inferred from the relevant equations so formulated that $dv/dx_\sharp = \infty$ and $K_v = \infty$ for all profiles at the channel bed (i.e., $v = 0$) in both sustaining and adverse channels, while $dv/dx_\sharp = \infty$ and K_v so evaluated at $v = 1$ for every profile should be relatively large, but not infinity, as illustrated in Fig. 5.1 for $N = 10/3$.
11. For gaining an insight into the effects of h_c/h_n and N on K_v at $v = 1$ at a glance, we use (5.29) and (5.30) to plot K_v at $v = 1$ against h_c/h_n with the N-value as a parameter, as shown in Fig. 5.11. Inspection of Fig. 5.11 reveals that K_v at $v = 1$ for GVF profiles in sustaining channels approaches infinity as $h_c/h_n \rightarrow 1$. This trend for K_v at $v = 1$ implies that the closer the two v-values at the two inflection points of the M profiles approach unity, as shown in Fig. 5.9, the larger is K_v at $v = 1$, irrespective of the M or S profile, as shown in Fig. 5.11. In contrast, K_v at $v = 1$ for A profiles in adverse channels decreases and approaches zero as $h_c/h_n \rightarrow \infty$. The A profile corresponding to this limit ($h_c/h_n = \infty$) is a vertical line with its K_v everywhere along the line is zero, as manifesting itself in Fig. 5.1.

References

Jan CD, Chen CL (2013) Gradually varied open-channel flow profiles normalized by critical depth and analytically solved by using gaussian hypergeometric functions. Hydrol Earth Syst Sci 17:973–987

Chow VT (1959) Open-Channel Hydraulics. McGraw-Hill, New York

Chen CL, Wang CT (1969) Nondimensional gradually varied flow profiles. J Hydraul Div ASCE 95(HY5):1671–1686

Chen CL, Chow VT (1971) Formulation of mathematical watershed-flow model. J Eng Mech Div ASCE 97(EM3):809–828

Appendix A
The Gaussian Hypergeometric Function and Its Application

A.1 Definition and Basic Relations of GHF

(1) For $| z | < 1$:

The Gaussian hypergeometric function, ${}_2F_1(a, b; c; z)$, which is symbolized herein as GHF for simplicity, can be expressed as an infinite series (see, e.g., Abramowitz and Stegun 1972, Korn and Korn 1961, Kummer 1836, Pearson 1974, and Luke 1975) as

$$ {}_2F_1(a, b; c; z) = \sum_{k=0}^{\infty} \left[\frac{(a)_k (b)_k}{(c)_k} \right] \frac{z^k}{k!}, \tag{A.1} $$

in which a, b, and c are the function parameters and z is the variable of the GHF. The parameters a, b, and c are independent of z and, in general, they may be complex. Such an infinite-series representation of the GHF in (A.1), is often referred to as the hypergeometric series; it is convergent for arbitrary a, b, and c (provided that c is neither a negative integer nor zero Seaborn (1991) and that a or b is not a negative integer (Whittaker and Watson, 1992) for real $-1 < z < 1$ (or $| z | < 1$), and for $z = \pm 1$ if $c > a+b$. The Pochhammer symbols $(a)_k$, $(b)_k$, and $(c)_k$ in the ascending factorial of a, b, and c are respectively defined in terms of the gamma function, $\Gamma(n)$, as

$$ (a)_k := \frac{\Gamma(a+k)}{\Gamma(a)} = \Pi_{n=0}^{k-1}(a+n), \tag{A.2} $$

$$ (b)_k := \frac{\Gamma(b+k)}{\Gamma(b)} = \Pi_{n=0}^{k-1}(b+n), \tag{A.3} $$

$$ (c)_k := \frac{\Gamma(c+k)}{\Gamma(c)} = \Pi_{n=0}^{k-1}(c+n). \tag{A.4} $$

Therefore, GHF can be written in the following form as

C.-D. Jan, *Gradually-varied Flow Profiles in Open Channels*,
Advances in Geophysical and Environmental Mechanics and Mathematics,
DOI: 10.1007/978-3-642-35242-3, © Springer-Verlag Berlin Heidelberg 2014

$$ {}_2F_1(a, b; c; z) = \frac{\Gamma(c)}{\Gamma(a)\Gamma(b)} \sum_{k=0}^{\infty} \frac{\Gamma(a+k)\Gamma(b+k)}{\Gamma(c+k)k!} z^k, \tag{A.5} $$

where the gamma function denoted by $\Gamma(n)$ for $n > 0$, is defined by

$$ \Gamma(n) = \int_0^{\infty} t^{n-1} e^{-t} dt \quad (\text{for } n > 0). \tag{A.6} $$

Using integration by parts, we will see that the gamma function satisfies the functional equation (i.e., the recurrence formula):

$$ \Gamma(n+1) = n\Gamma(n). \tag{A.7} $$

If n is a positive integer, $\Gamma(n+1) = n!$. For example, we can generalize the gamma function to $n < 0$ by using (A.7) in the form as

$$ \Gamma(n) = \frac{\Gamma(n+1)}{n} \quad (\text{for } n < 0). \tag{A.8} $$

$\Gamma(n)$ is convergent for $n > 0$ and $n < 0$ except when n is a non-positive integer. Based on (A.6) and (A.8), we can plot the variation of $\Gamma(n)$ against n. Figure A.1 shows the gamma function $\Gamma(n)$ along part of the real axis. This figure was gererated by using a plot command of the Mathematica software. Some particular values of the gamma function are: $\Gamma(-1/2) = -2\sqrt{\pi}$, $\Gamma(0^+) = \infty$, $\Gamma(0^-) = -\infty$, $\Gamma(1/2) = \sqrt{\pi}$, $\Gamma(1) = 1$, $\Gamma(3/2) = \sqrt{\pi}/2$, $\Gamma(2) = 1$, and $\Gamma(5) = 4! = 24$.

As for the commutative rule of the GHF, it can be readily seen from the hypergeometric series in (A.1) that the parameters, a and b, in the GHF are commutative; therefore,

$$ {}_2F_1(a, b; c; z) = {}_2F_1(b, a; c; z). \tag{A.9} $$

Equation (A.9) shows a commutative rule in the GHF, connoting that the parameters, b and a, in the GHF-based solution are independent of the order in the operation of multiplication within each term in the hypergeometric series (A.1), thus being commutative with each other. It merits attention to distinguish the expression of ${}_2F_1(a, b; c; z)$ in the Mathematica version 7.0.1 [2009] from that of ${}_2F_1(b, a; c; z)$ in its earlier versions, including version 7 [2008]. The reverse order of the two parameters, b and a, in the expression of the GHF in the earlier versions actually accounts for the commutative rule of the GHF, as represented by (A.9). In truth, both GHFs in (A.9) expressed using the hypergeometric series, (A.1), are exactly identical.

(2) For $|z| > 1$ **:**

A survey of the literature reveals that there are two linearly independent solutions of the hypergeometric differential equation at each of the three singular points $z =$ 0, 1, and ∞ for a total of six special solutions, which are in fact fundamental to Kummer's (1836) 24 solutions. Besides one of two linearly independent solutions

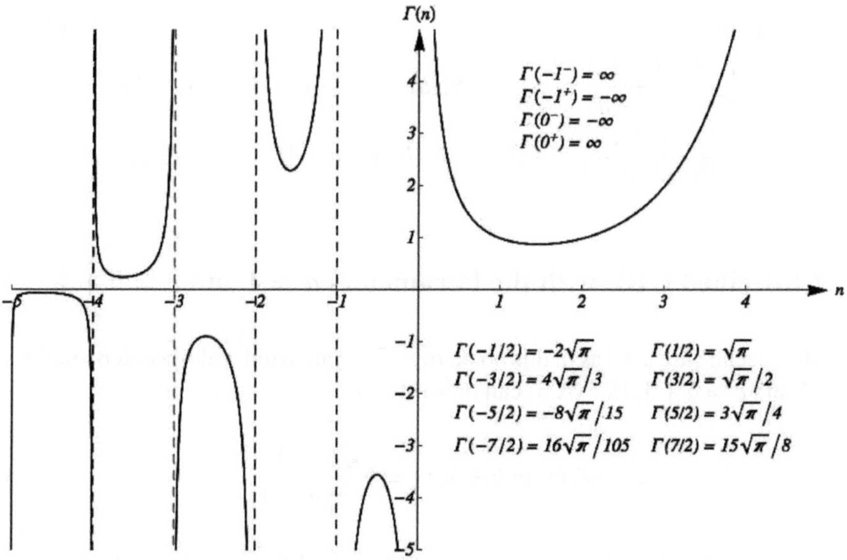

Fig. A.1 The gamma function $\Gamma(n)$ along part of the real axis. $\Gamma(n)$ is convergent for $n > 0$ and $n < 0$ except n = non-positive integers. The figure was generated by the plot command of the Mathematica software: Plot [Gamma[x], {x, –5, 5}, AxesStyle → Arrowheads [{0.03, 0.03}], PlotRange → {{-5, 5}, {-5, 5}}, BaseStyle → {FontWeight → "Bold", FontSize → 12}]

obtained around $z = 0$ is the GHF, as given in (A.1), we have a pair of linearly independent solutions around $z = \infty$ as follows: For $|\, z \,| > 1$ (i.e., covering both domains $z < -1$ and $z > 1$), if $(a - b)$ is not an integer,

$$(-z^{-1})^a_2 F_1\,(\,a, a - c + 1; a - b + 1; z^{-1}\,) \qquad \text{(A.10)}$$
$$= (-z^{-1})^a \sum_{k=0}^{\infty} \left[\frac{(a)_k (a - c + 1)_k}{(a - b + 1)_k} \right] \frac{z^{-k}}{k!},$$

$$(-z^{-1})^b_2 F_1(b, b - c + 1; -a + b + 1; z^{-1}) \qquad \text{(A.11)}$$
$$= (-z^{-1})^a \sum_{k=0}^{\infty} \left[\frac{(b)_k (b - c + 1)_k}{(-a + b + 1)_k} \right] \frac{z^{-k}}{k!}.$$

Any three of the aforementioned six special solutions satisfy a relation, thus giving rise to twenty combinations thereof. Among them, there is a relation connecting one GHF in the domain of $|\, z \,| < 1$ to two GHF in the domain of $|\, z \,| > 1$ in the following way (Luke 1975):

$$
\begin{aligned}
&{}_2F_1(a, b; c; z) \\
&= \frac{\Gamma(b-a)\Gamma(c)}{\Gamma(b)\Gamma(c-a)} (-z^{-1})^a \, {}_2F_1(a, a-c+1; a-b+1; z^{-1}) \\
&+ \frac{\Gamma(a-b)\Gamma(c)}{\Gamma(a)\Gamma(c-b)} (-z^{-1})^b \, {}_2F_1(b, b-c+1; -a+b+1; z^{-1}).
\end{aligned} \tag{A.12}
$$

A.2 Specified GHF with the Parameters $a = 1$ and $c = b + 1$

For the special case used as in this book, a, b, and c are fixed with specified relations, $a = 1$ and $c = b + 1$, Eq. (A.5) can be written as

$$
{}_2F_1(1, b; b+1, z) = b \sum_{k=0}^{\infty} \frac{z^k}{b+k}. \tag{A.13}
$$

The GHF used in the solutions of the gradually-varied-flow profiles in this book are all in the form of Eq. (A.13), as mentioned in Chap. 3. Since the first argument of the GHF of Eq. (A.13) is always unity, the second and third arguments differ in one unity, and the fourth argument is a variable only, we can express the GHF in a simpler expression as shown in the following equation for reducing the related equations to shorter expressions for facilitating the reading of the manuscript, as treated by Jan and Chen (2012):

$$
g(b, z) = {}_2F_1(1, b; b+1, z) = b \sum_{k=0}^{\infty} \frac{z^k}{b+k}. \tag{A.14}
$$

Therefore, for shortening expressions, we use $g(b, z)$ instead of $F(1, b; b+1; z)$ to represent GHF when we treat the GHF-based GVF solutions.

A.3 Some Examples of Specified GHF

Figure A.2 shows four examples of Gaussian hypergeometric functions, $g(2/5, u^{10/3})$ and $g(-3/5, u^{10/3})$ for $0 < u < 1$, and $g(3/5, u^{-10/3})$ and $g(-3/10, u^{-10/3})$ for $1 < u < 2$. Figure A.3 shows another examples of Gaussian hypergeometric functions, $g(1/N, u^N)$ and $g(-1/N, u^N)$ for $0 < u < 1$, and $g(1/N, u^{-N})$ and $g(-1/N, u^{-N})$ for $1 < u < 2$, with four different values of hydraulic exponent N varying from 2 to 5. These figures show that $g(b, z) \to 1$ as $z \to 0$, $g(b, z) \to -\infty$ as b is negative and $z \to 1$, and $g(b, z) \to \infty$ as b is positive and $z \to 1$, in which z ($= u^N$ or u^{-N}) is real variable of positive values. On the other hand, as z ($= -u^N$) is real variable of negative values, $g(b, z) \to$ a finite value as $z \to -1$ (i.e.,$u \to 1$), as

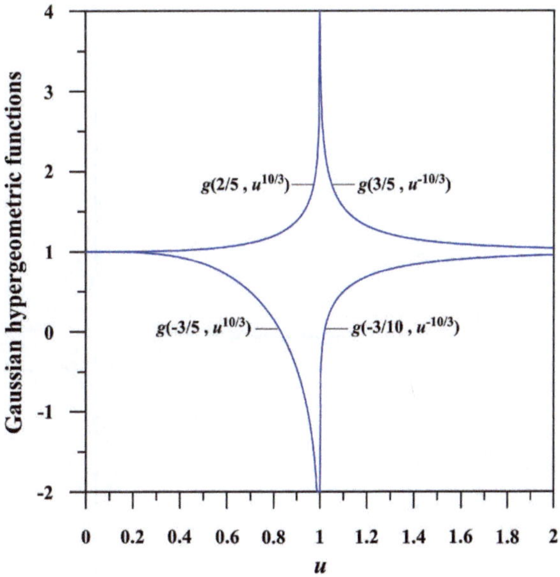

Fig. A.2 Four examples of the GHF having positive variables. These GHF diverge at $u = 1$, and converge to unity as u approachs either zero or the infinite. This type of GHF could be used in the GHF-based GVF solutions in sustaining channels

shown in Fig. A.4. Figure A.5 shows another examples of Gaussian hypergeometric functions, $g(1/N, -u^N)$ and $g(-1/N, -u^N)$ for $0 < u < 1$, and $g(1/N, -u^N)$ and $g(-1/N, -u^N)$ for $1 < u < 2$, with four different values of hydraulic exponent N varying from 2 to 5. These figures show that $g(b, z) \to 1$ as $z \to 0$, and $g(b, z) \to$ a finite value as $z \to -1$ (i.e., $u \to 1$, as $z = -u^N$ or $z = -u^{-N}$). These kinds of Gaussian hypergeometric functions, $g(b, -u^N)$ and $g(b, -u^{-N})$, will be used to express the solutions of the surface profiles of gradually-varied flow in channels of an adverse slope (a negative slope). It means that there is no singularity for the surface slope of the profiles of gradually-varied flow in adverse channels.

A.4 Integral Solutions in Terms of GHF

For example, the gradually-varied flow (GVF) in sustaining channels has a positive variable $z = u^N$, for the case of $0 \leq u < 1$, the solution of the indefinite integral in the GVF equation can be expressed in terms of GHF as

$$\int \frac{u^{\phi}}{1 - u^N} du = \frac{u^{\phi+1}}{\phi + 1} g\left(\frac{\phi + 1}{N}, u^N\right) + \text{Const.,} \tag{A.15}$$

while, for the case of $u > 1$, letting $w = u^{-1}$, the solution of the indefinite integral in terms of GHF is written as

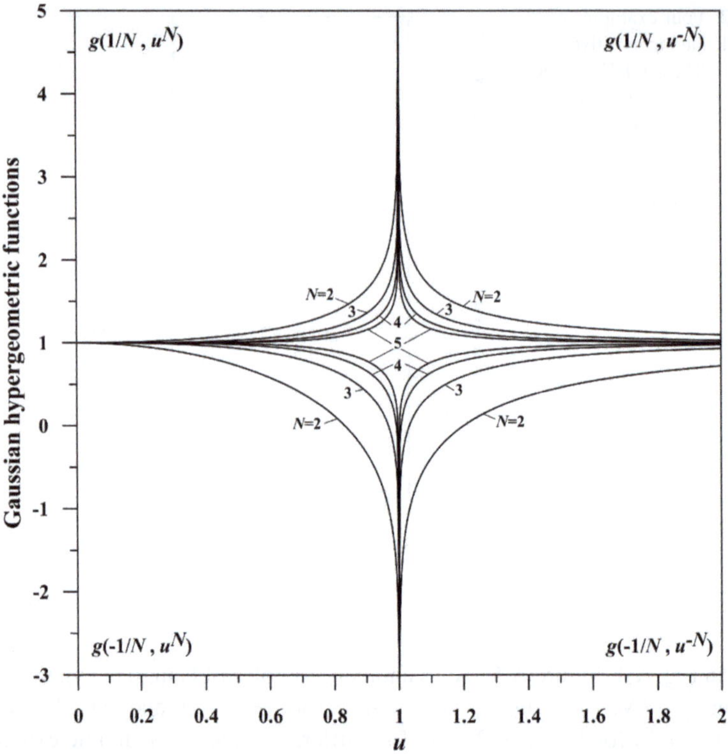

Fig. A.3 Effect of the hydraulic exponent N on the variation of GHF having positive variables. These GHF diverge at $u = 1$, and converge to unity as u approaches either zero or the infinte

Fig. A.4 Four examples of the GHF having negative variables. These GHF do not diverge at $u = 1$. This type of GHF can be used in the GHF-based GVF solutions in adverse channels

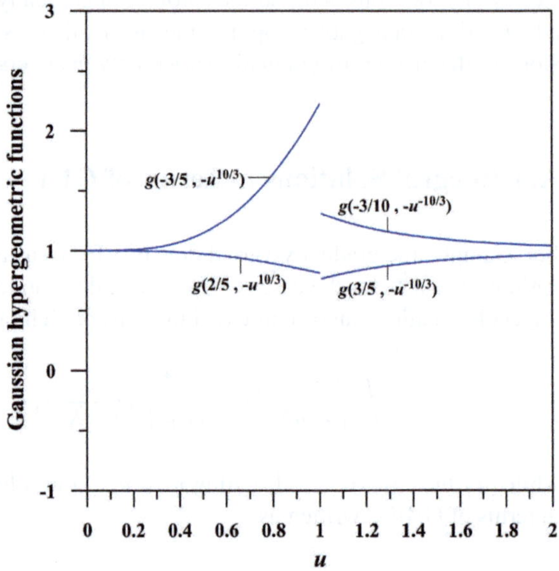

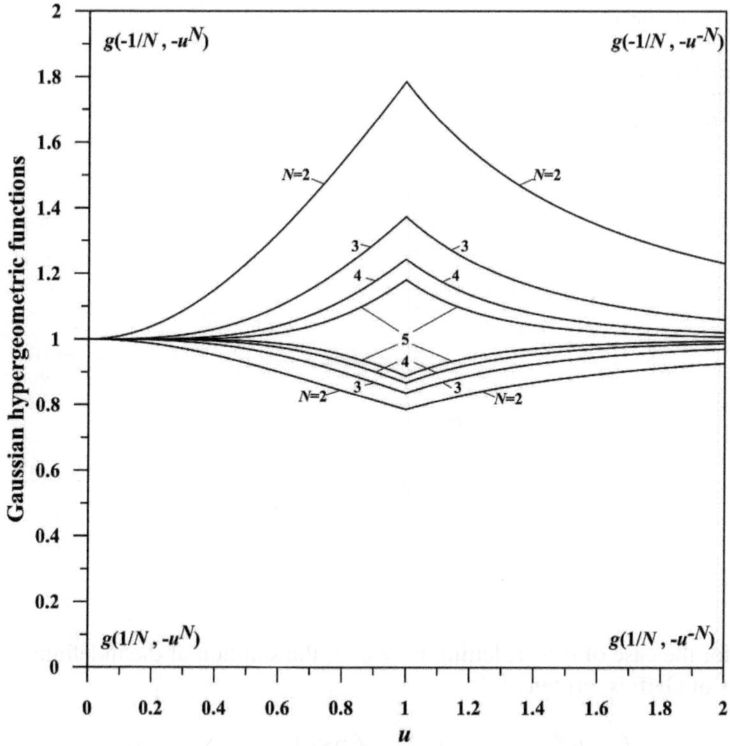

Fig. A.5 Effect of the hydraulic exponent N on the variation of GHF having negative variables. These GHF do not diverge at $u = 1$, and converge to unity as u approaches either zero or the infinte

$$\int \frac{w^{\phi}}{1 - w^N} dw = \frac{w^{\phi+1}}{\phi + 1} g\left(\frac{\phi + 1}{N}, w^N\right) + \text{Const.}. \tag{A.16}$$

(A.16), on substitution of $w = u^{-1}$, yields

$$\frac{w^{(\phi+1)}}{\phi + 1} g\left(\frac{\phi + 1}{N}, w^N\right) = \frac{u^{-(\phi+1)}}{\phi + 1} g\left(\frac{\phi + 1}{N}, u^{-N}\right), \tag{A.17}$$

in which the parameters ϕ and N are real numbers.

On the other hand, the gradually-varied flow (GVF) in adverse channels has a negative slope and negative varible $z = -u^N$, for the case of $0 \leq u < 1$, the solution of the indefinite integral in the GVF equation for flow in adverse channels can be expressed in terms of GHF as

$$\int \frac{u^{\phi}}{1 + u^N} du = \frac{u^{\phi+1}}{\phi + 1} g\left(\frac{\phi + 1}{N}, -u^N\right) + \text{Const.}, \tag{A.18}$$

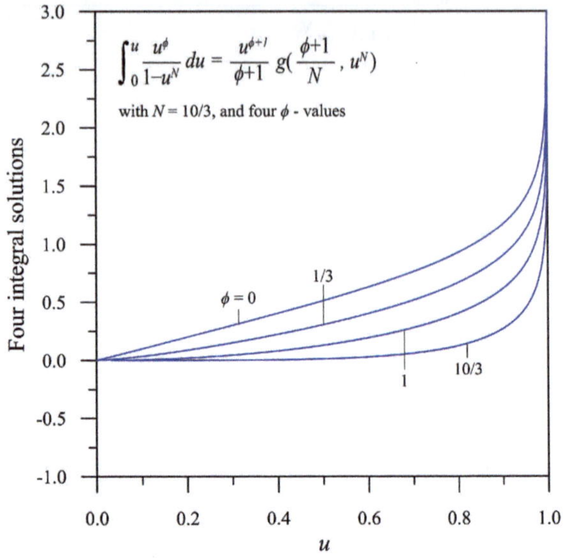

Fig. A.6 Four examples of the integral solutions in terms of GHF for the specified case of $N = 10/3$, and $\phi = 0, 1/3$, 1.0, and 10/3, respectively

while, for the case of $u > 1$, letting $w = u^{-1}$, the solution of the indefinite integral in terms of GHF is written as

$$\int \frac{w^\phi}{1+w^N} dw = \frac{w^{\phi+1}}{\phi+1} g\left(\frac{\phi+1}{N}, -w^N\right) + \text{Const.}. \tag{A.19}$$

Equation (A.19), on substitution of $w = u^{-1}$, yields

$$\frac{w^{\phi+1}}{\phi+1} g\left(\frac{\phi+1}{N}, -w^N\right) = \frac{u^{-(\phi+1)}}{\phi+1} g\left(\frac{\phi+1}{N}, -u^{-N}\right). \tag{A.20}$$

Figure A.6 shows four examples of the integral solutions in terms of GHF for the specified case of $N = 10/3$, and $\phi = 0$, 1/3, 1.0 and 10/3, respectively. These integral solutions show that the results of integration diverge as $u \to 1$,

Appendix B
Proof of an Identity Relation Between Two Contiguous GHF

According to (A.1) or (A.5), the Gaussian hypergeometric function (GHF) can be written in the form

$$ {}_2F_1(a, b; c; z) = \sum_{k=0}^{\infty} \frac{\Gamma(a+k)}{\Gamma(a)} \frac{\Gamma(b+k)}{\Gamma(b)} \frac{\Gamma(c+k)}{\Gamma(c)} \frac{z^k}{k!}. \tag{B.1} $$

If the parameters a, b, and c in the GHF are fixed with specified relations, $a = 1$ and $c = b + 1$ as used in this book, the following functional identity relation between the two contiguous GHF can be generally established as

$$ {}_2F_1(1, b; b+1, z) = 1 + \frac{bz}{b+1}\, {}_2F_1(1, b+1; b+2, z). \tag{B.2} $$

By means of the induction method, we can validate (B.2) with the help of (A.1) through (A.4) as follows:

$$ \begin{aligned} & 1 + \frac{bz}{b+1}\, {}_2F_1(1, b+1; b+2; z) \\ & = 1 + \frac{bz}{b+1} \sum_{k=0}^{\infty} \left[\frac{\Gamma(1+k)}{\Gamma(1)} \frac{\Gamma(b+1+k)}{\Gamma(b+1)} \frac{\Gamma(b+2)}{\Gamma(b+2+k)} \right] \frac{z^k}{k!} \\ & = 1 + \frac{b}{b+1} \sum_{k=0}^{\infty} \left[\frac{\Gamma(1+k)}{\Gamma(1)} \frac{\Gamma(b+1+k)}{b\Gamma(b)} \frac{(b+1)\Gamma(b+1)}{\Gamma(b+2+k)} \right] \frac{z^{k+1}}{k!} \\ & \text{(letting } j = k+1) \\ & = 1 + \sum_{j=1}^{\infty} \left[\frac{j\Gamma(j)}{\Gamma(1)} \frac{\Gamma(b+j)}{\Gamma(b)} \frac{\Gamma(b+1)}{\Gamma(b+1+j)} \right] \frac{z^j}{j(j-1)!} \end{aligned} \tag{B.3} $$

C.-D. Jan, *Gradually-varied Flow Profiles in Open Channels*,
Advances in Geophysical and Environmental Mechanics and Mathematics,
DOI: 10.1007/978-3-642-35242-3, © Springer-Verlag Berlin Heidelberg 2014

$$= \sum_{j=0}^{\infty} \left[\frac{\Gamma(1+j)}{\Gamma(1)} \frac{\Gamma(b+j)}{\Gamma(b)} \frac{\Gamma(b+1)}{\Gamma(b+1+j)} \right] \frac{z^j}{j!}$$

$$= {}_2F_1(1, b; b+1; z).$$

The functional identity relation between two contiguous GHF as proved in the derivation process of (B.3) is also called a recurrence formulas. Using a simplified expression of GHF, i.e., $g(b, z)$, to replace ${}_2F_1(1, b; b+1; z)$ as shown in (A.14) in Appendix A, we can rewrite the recurrence formulas into a shorter expression as done by Jan and Chen (2012, 2013).

$$g(b, z) = 1 + \frac{bz}{b+1} g(b+1, z). \tag{B.4}$$

The so-called recurrence formulas, such as (3.32), (3.33), (3.45), (3.46), (3.47) and (3.48) in Chap. 3, and (4.35), (4.36), (4.45), (4.46), (4.50), (4.51), (4.72), (4.73), (4.82), (4.83), (4.87) and (4.88) in Chap. 4, can be proved to possess the property of the functional identity relation between two contiguous GHF, as shown in (B.2) or (B.4). For convenience of discussion, we take the example of (3.33) in Chap. 3, listed here as

$$g\left(\frac{-M+1}{N}, u^N\right) = 1 + \frac{(-M+1)u^N}{N-M+1} g\left(\frac{N-M+1}{N}, u^N\right). \tag{B.5}$$

It can be readily identified from the two contiguous GHF in (3.33) in that the parameters and variable in the GHF on its left-hand side of the equal sign is constituted by $a = 1$, $b = (-M+1)/N$, $c = (N-M+1)/N$, and $z = u^N$, thus satisfying one of the two parameter relations, $a = 1$ and $c = b + 1$, while the other GHF on its right-hand side of the equal sign has the other parameter relations, $a = 1$, $b = (N-M+1)/N$, and $c = (2N-M+1)/N$, thus meeting all the parameter relations specified on both sides of (B.2) or (B.4). For the specified case of $M = 3$ and $N = 10/3$, (B.5) [i.e., (3.33)] yields

$$g\left(\frac{-3}{5}, u^{10/3}\right) = 1 - \frac{3}{2} u^{10/3} g\left(\frac{2}{5}, u^{10/3}\right). \tag{B.6}$$

Similarly, we can prove that each of the above-mentioned recurrence formulas in Chaps. 3 and 4 has the same parameter relations in such two contiguous GHF, thus satisfying (B.4).

Appendix C
Obtain (3.43) Directly from (3.30) by Using the Transformation Relation (A.12), and Vice Versa

We start with the GHF-based solutions of (3.8) in Chap. 3 in the domains $|\ u\ | < 1$ and $|\ u\ | > 1$, as shown in (3.30) and (3.43), respectively,

$$x_* = u\left[1 - g\left(\frac{1}{N}, u^N\right)\right] + \frac{\lambda^M u^{N-M+1}}{N-M+1} g\left(\frac{N-M+1}{N}, u^N\right) + \text{Const.}, \tag{3.30}$$

$$x_* = u\left[1 - \frac{u^{-N}}{N-1} g\left(\frac{N-1}{N}, u^{-N}\right)\right] + \frac{\lambda^M u^{-M+1}}{M-1} g\left(\frac{M-1}{N}, u^{-N}\right) + \text{Const.}. \tag{3.43}$$

Equation (A.12) in Appendix A has shown that there is a relation connecing one GHF in the domain of $|\ z\ | < 1$ to two GHF in the domain of $|\ z\ | > 1$ (Luke 1975). For the specified parameters $a = 1$ and $c = b + 1$, (A.12) connecting one GHF in the domain of $|\ z\ | < 1$ to two GHF in the domain of $|\ z\ | > 1$ can be simplified as

$$\begin{aligned} {}_2F_1(1, b; 1+b; z) &= \frac{\Gamma(b-1)\Gamma(b+1)}{\Gamma(b)\Gamma(b)} (-z^{-1})^1\ {}_2F_1(1, 1-b; 2-b; z^{-1}) \\ &+ \frac{\Gamma(1-b)\Gamma(1+b)}{\Gamma(1)\Gamma(1)} (-z^{-1})^b\ {}_2F_1(b, 0; b; z^{-1}). \end{aligned} \tag{C.1}$$

Since $\Gamma(b) = (b-1)\Gamma(b-1)$, $\Gamma(b+1) = b\Gamma(b)$, and ${}_2F_1(b, 0; b; z^{-1}) = 1$, Eq. (C.1) yields

$$\begin{aligned} {}_2F_1(1, b; 1+b; z) = \frac{b}{b-1}(-z^{-1})\ {}_2F_1(1, 1-b; 2-b; z^{-1}) \\ + \Gamma(1-b)\Gamma(1+b)(-z^{-1})^{-b}. \end{aligned} \tag{C.2}$$

C.-D. Jan, *Gradually-varied Flow Profiles in Open Channels*,
Advances in Geophysical and Environmental Mechanics and Mathematics,
DOI: 10.1007/978-3-642-35242-3, © Springer-Verlag Berlin Heidelberg 2014

For reducing the related equations to shorter expressions for facilitating the reading of the manuscript, we will use $g(b, z)$ instead of ${}_2F_1(1, b; b+1; z)$ in this book to represent GHF, namely, $g(b, z) = {}_2F_1(1, b; b+1; z)$. Therefore, the shorter expression of (C.2) is given as

$$g(b, z) = \frac{b}{b-1}(-z^{-1})g(1-b, z^{-1}) + \Gamma(1-b)\Gamma(1+b)(-z^{-1})^{-b}. \qquad \text{(C.3)}$$

One may use (C.3) to transform the GHF-based solutions of (3.8) in the domain of $|u| < 1$ to their counterparts in the domain of $|u| > 1$ without recourse to the formulation of (3.14) [i.e., an alternative form of (3.8) with its variable, w, being expressed as u^{-1}], which can be solved for GVF profiles in the domain of $|u^{-1}| < 1$ (or $|u| > 1$). In this Appendix, we focus our attention on use of (C.3) to transform (3.30)– (3.43), thereby proving that we can obtain (3.43) directly from (3.30) rather than indirectly through (3.14).

Before using (C.3) to find the GHF-based solutions of (3.8) for the entire domain of u, namely $0 \leq u < \infty$, excluding at $u = 1$ attention is focused on their generality in that the relation, (C.3), which originally connects one GHF in the domain of $|z| < 1$ to two GHF in the domain of $|z| > 1$, can also be applied in reverse, i.e., one GHF in the domain of $|z| > 1$ being transformed into two GHF in the domain of $|z| < 1$. In particular, for the specified parameter relations, $a = 1$ and $c = b + 1$, we obtain the reverse relation from (A.12) as follows: Substituting $(1 - b)$ for b and z for z^{-1} into both sides of (C.3) and then rearranging its result through the relations $\Gamma(b) = \Gamma(1 + b)/b$ and $\Gamma(2 - b) = (1 - b)\Gamma(1 - b)$ yields

$$\begin{aligned} g(1-b, z^{-1}) = & \frac{b-1}{b}(-z)g(b, z) \\ & - \frac{b-1}{b}\Gamma(1+b)\Gamma(1-b)(-z)^{1-b}. \end{aligned} \qquad \text{(C.4)}$$

In fact, we can derive (C.4) from (C.3) directly by means of transposition. The reversibility of (C.3) for the GHF to transform its domain of validity from $|z| < 1$ to $|z| > 1$, or vice versa, namely (C.4) for the GHF to switch its valid domain of $|z|$ in reverse makes such relations instrumental when the GHF-based solutions are used in the computation of GVF profiles. Such an exclusive property of the GHF in effect makes the GHF-based solutions of the dimensionless GVF Eq. (3.8) versatile in application. A few aspects of such relations proved in the GHF-based solutions are given below.

C.1 Obtain (3.43) Directly from (3.30) by Using the Transformation Relation

For example, the relation connecting one GHF in the domain of $| z | < 1$ to two GHF in the domain of $| z | > 1$, as shown in (C.3), can be applied to transform the GHF-based solution of (3.8) in the domain of $| u | < 1$, namely (3.30), to the corresponding GHF-based solution of (3.8) in the domain of $| u | > 1$. Using (C.3), we can prove that (3.30) in the domain of $| u | < 1$ is transformed to (3.43) in the domain of $| u | > 1$, though (3.43) has been actually obtained from (3.14) rather than from (3.8). Because we have $b = 1/N$ and $b = (N - M + 1)/N$ in the first and second GHF with $z = u^N$ appearing on the right-hand side of (3.30), respectively, (C.3) on substitution of such respective b-and z-expressions yields the following two relations

$$g\left(\frac{1}{N}, u^N\right) = \frac{u^{-N}}{N-1} g\left(\frac{N-1}{N}, u^{-N}\right) + \Gamma\left(\frac{N-1}{N}\right)\Gamma\left(\frac{N+1}{N}\right)(-u^{-N})^{1/N}, \tag{C.5}$$

$$g\left(\frac{N-M+1}{N}, u^N\right) = \frac{(N-M+1)u^{-N}}{M-1} g\left(\frac{M-1}{N}, u^{-N}\right) + \Gamma\left(\frac{M-1}{N}\right)\Gamma\left(\frac{2N-M+1}{N}\right)(-u^{-N})^{(N-M+1)/N}. \tag{C.6}$$

Substituting the first and second GHF appearing on the right-hand side of (3.30) from (C.5) and (C.6), respectively, into (3.30) yields

$$\begin{aligned}
x_* &= u\left[1 - g\left(\frac{1}{N}, u^N\right)\right] + \frac{\lambda^M u^{N-M+1}}{N-M+1} g\left(\frac{N-M+1}{N}, u^N\right) + \text{Const.} \\
&= u\left[1 - \frac{u^{-N}}{N-1} g\left(\frac{N-1}{N}, u^{-N}\right) - \Gamma\left(\frac{N-1}{N}\right)\Gamma\left(\frac{N+1}{N}\right)(-u^{-N})^{1/N}\right] \\
&+ \frac{\lambda^M u^{N-M+1}}{N-M+1}\left[\frac{(N-M+1)u^{-N}}{M-1} g\left(\frac{M-1}{N}, u^{-N}\right)\right. \\
&\left.+ \Gamma\left(\frac{M-1}{N}\right)\Gamma\left(\frac{2N-M+1}{N}(-u^{-N})^{(N-M+1)/N}\right)\right] + \text{Const.} \\
&= u\left[1 - \frac{u^{-N}}{N-1} g\left(\frac{N-1}{N}, u^{-N}\right)\right] + \frac{\lambda^M u^{-M+1}}{M-1} g\left(\frac{M-1}{N}, u^{-N}\right) \\
&- (-1)^{1/N}\Gamma\left(\frac{N-1}{N}\right)\Gamma\left(\frac{N+1}{N}\right) \\
&+ \frac{\lambda^M(-1)^{(N-M+1)/N}}{N-M+1}\Gamma\left(\frac{M-1}{N}\right)\Gamma\left(\frac{2N-M+1}{N}\right) + \text{Const.}
\end{aligned}$$

$$
\begin{aligned}
&= u\left[1 - \frac{u^{-N}}{N-1} g\left(\frac{N-1}{N}, u^{-N}\right)\right] \\
&+ \frac{\lambda^M u^{-M+1}}{M-1} g\left(\frac{M-1}{N}, u^{-N}\right) + \text{Const}^*.
\end{aligned}
\tag{C.7}
$$

It merits attention to state that (C.7) is identical to (3.13) because "Const*" in (C.7) that contains two constant terms appearing in (C.7) in addition to "Const." does not lose its generality in comparison. Consequently, one may conclude that (3.43) obtained from (3.14) is indeed the GHF-based solution of (3.8) in the domain of $| u | > 1$. This reconfirms the validity of (3.43) in the domain of $| u | > 1$ and the uniqueness of the role of (3.43) playing in the computation of GVF profiles in sustaining channels.

C.2 Obtain (3.30) Directly from (3.43) by Using the Transformation Relation

The application of (C.4) in the reverse transformation of one GHF in the domain of $| z | > 1$ to two GHF in the domain of $| z | < 1$ is exemplified when (3.43) in the domain of $| u | > 1$ is transformed to (3.30) in the domain of $| u | < 1$. In this case, because we again have $b = 1/N$ and $b = (N - M - 1)/N$ in the first and second GHF with $z = u^{-N}$ appearing on the right-hand side of (3.43), respectively, (C.4) on substitution of such respective b- and z-expressions yields the following two relations

$$
\begin{aligned}
g\left(\frac{N-1}{N}, u^{-N}\right) &= (N-1)u^N g\left(\frac{1}{N}, u^N\right) \\
&+ (N-1)\Gamma\left(\frac{N+1}{N}\right)\Gamma\left(\frac{N-1}{N}\right)(-u^N)^{1-1/N},
\end{aligned}
\tag{C.8}
$$

$$
\begin{aligned}
g\left(\frac{M-1}{N}, u^{-N}\right) &= \frac{(M-1)u^N}{N-M+1} g\left(\frac{N-M+1}{N}, u^N\right) \\
&+ \frac{(M-1)(-u^N)^{(M-1)/N}}{N-M+1}\Gamma\left(\frac{2N-M+1}{N}\right)\Gamma\left(\frac{M-1}{N}\right).
\end{aligned}
\tag{C.9}
$$

Substituting the first and second GHF appearing on the right-hand side of (3.43) from (C.8) and (C.9), respectively, into (3.43) yields

$$\begin{aligned}
x_* &= u\left[1 - \frac{u^{-N}}{N-1} g\left(\frac{N-1}{N}, u^{-N}\right)\right] \\
&\quad + \frac{\lambda^M u^{-M+1}}{M-1} g\left(\frac{M-1}{N}, u^{-N}\right) + \text{Const.} \\
&= u\left[1 - g\left(\frac{1}{N}, u^N\right) - u^{-N}\Gamma\left(\frac{N+1}{N}\right)\Gamma\left(\frac{N-1}{N}\right)(-u^N)^{1-1/N}\right] \\
&\quad + \frac{\lambda^M u^{-M+1}}{M-1}\left[\frac{(M-1)u^N}{N-M+1} g\left(\frac{N-M+1}{N}, u^N\right)\right] \\
&\quad + \frac{\lambda^M u^{-M+1}}{M-1}\left[\frac{(M-1)(-u^N)^{(M-1)/N}}{N-M+1}\right. \\
&\qquad \left. \times \Gamma\left(\frac{2N-M+1}{N}\right)\Gamma\left(\frac{M-1}{N}\right)\right] + \text{Const.} \\
&= u\left[1 - g\left(\frac{1}{N}, u^N\right)\right] + \frac{\lambda^M u^{N-M+1}}{N-M+1} g\left(\frac{N-M+1}{N}, u^N\right) \\
&\quad + (-1)^{-1/N}\Gamma\left(\frac{N+1}{N}\right)\Gamma\left(\frac{N-1}{N}\right) \\
&\quad + \frac{\lambda^M (-1)^{(M-1)/N}}{N-M+1}\Gamma\left(\frac{2N-M+1}{N}\right)\Gamma\left(\frac{M-1}{N}\right) + \text{Const.} \\
&= u\left[1 - g\left(\frac{1}{N}, u^N\right)\right] + \frac{\lambda^M u^{N-M+1}}{N-M+1} g\left(\frac{N-M+1}{N}, u^N\right) + \text{Const}^*.
\end{aligned} \tag{C.10}$$

It is noted that (C.10) is identical to (3.30) because "Const*" in (C.10) that contains two constant terms appearing in (C.10) in addition to "Const." does not lose its generality in comparison.

Though we have proved that (3.30) can be transformed to (3.43), or vice versa, using a couple of the relations connecting one GHF in the domain of $|u| < 1$ to two GHFs in the domain of $|u| > 1$, or vice versa in the reverse transformation, we can by no means claim that (3.30) and (3.43) are identical because they are only valid and convergent in their respective domains of u. In other words, (3.30) is only valid in the domain of $|u| < 1$, whereas (3.43) is only valid in the domain of $|u| > 1$. Consequently, suffice it to say that the complete GHF-based solutions of (3.8) are constituted by a combination of (3.30) and (3.43).

Appendix D
ETF-Based Solutions of the First and Second Integrals in (3.12)

Let us start with (3.12) in Chap. 3 that is a direct integral solution of (3.8) for GVF in sustaining channels:

$$x_* = u - \int_0^u \frac{1}{1-u^N}du + \lambda^M \int_0^u \frac{u^{N-M}}{1-u^N}du + \text{Const.}. \tag{3.12}$$

With the help of the Mathematica software (Wolfram 1996) or the Maple software (Bernardin et al. 2011), the alternative analytical solutions of the first and second integrals appearing in the above equation for $N = 3$, 10/3, 7/2, 17/5 and 11/3 for fully rough flows in wide channels ($M = 3$) can be expressed in terms of elementary transcendental functions (ETF). The results are shown below.

D.1 ETF-Based Solutions of the First Integral for Five N-Values

$$(1)\quad \int_0^u \frac{1}{1-u^3}du \quad (\text{for } N = 3) \tag{D.1}$$
$$= \frac{1}{\sqrt{3}}\arctan\left[\frac{1+2u}{\sqrt{3}}\right] - \frac{1}{3}\ln[-1+u] + \frac{1}{6}\ln[1+u+u^2].$$

$$(2)\quad \int_0^u \frac{1}{1-u^{10/3}}du \quad (\text{for } N = 10/3) \tag{D.2}$$
$$= \frac{3}{40}\left\{-2\sqrt{2(5+\sqrt{5})}\arctan\left[\frac{1+\sqrt{5}-4u^{1/3}}{\sqrt{10-2\sqrt{5}}}\right]\right.$$
$$-2\sqrt{10-2\sqrt{5}}\arctan\left[\frac{1-\sqrt{5}+4u^{1/3}}{\sqrt{2(5+\sqrt{5})}}\right]$$

C.-D. Jan, *Gradually-varied Flow Profiles in Open Channels*,
Advances in Geophysical and Environmental Mechanics and Mathematics,
DOI: 10.1007/978-3-642-35242-3, © Springer-Verlag Berlin Heidelberg 2014

$$
\begin{aligned}
&- 2\sqrt{10-2\sqrt{5}}\arctan\left[\frac{-1+\sqrt{5}+4u^{1/3}}{\sqrt{2(5+\sqrt{5})}}\right]\\
&+ 2\sqrt{10-2\sqrt{5}}\arctan\left[\frac{1+\sqrt{5}+4u^{1/3}}{\sqrt{10-2\sqrt{5}}}\right]\\
&- 4\ln\left[-1+u^{1/3}\right]+4\ln\left[1+u^{1/3}\right]\\
&+ (1+\sqrt{5})\ln\left[1-\frac{1}{2}(-1+\sqrt{5})u^{1/3}+u^{2/3}\right]\\
&- (1+\sqrt{5})\ln\left[1+\frac{1}{2}(-1+\sqrt{5})u^{1/3}+u^{2/3}\right]\\
&+ (-1+\sqrt{5})\ln\left[1-\frac{1}{2}(1+\sqrt{5})u^{1/3}+u^{2/3}\right]\\
&- (-1+\sqrt{5})\ln\left[1+\frac{1}{2}(1+\sqrt{5})u^{1/3}+u^{2/3}\right]\Bigg]\Bigg\}.
\end{aligned}
$$

$$
\begin{aligned}
(3)\quad &\int_0^u \frac{1}{1-u^{7/2}}du \quad (\text{for } N = 7/2) \qquad \text{(D.3)}\\
&= \frac{4}{7}\arctan\left[\sec[\frac{3\pi}{14}](\sqrt{u}-\sin[\frac{3\pi}{14}])\right]\cos[\frac{\pi}{14}]\\
&- \frac{4}{7}\arctan\left[\csc[\frac{\pi}{7}](\sqrt{u}+\cos[\frac{\pi}{7}])\right]\cos[\frac{3\pi}{14}]\\
&- \frac{2}{7}\ln[-1+\sqrt{u}]+\frac{2}{7}\cos[\frac{\pi}{7}]\ln\left[1+u+2\sqrt{u}\sin[\frac{\pi}{14}]\right]\\
&+ \frac{2}{7}\sin[\frac{\pi}{14}]\ln\left[1+u-2\sqrt{u}\sin[\frac{3\pi}{14}]\right]\\
&- \frac{4}{7}\arctan\left[\sec[\frac{\pi}{14}](\sqrt{u}+\sin[\frac{\pi}{7}])\right]\sin[\frac{\pi}{7}]\\
&- \frac{2}{7}\sin[\frac{3\pi}{14}]\ln\left[1+u+2\sqrt{u}\cos[\frac{\pi}{7}]\right].
\end{aligned}
$$

$$
\begin{aligned}
(4)\quad &\int_0^u \frac{1}{1-u^{17/5}}du \quad (\text{for } N = 17/5) \qquad \text{(D.4)}\\
&= -\frac{10}{17}\arctan\left[\sec[\frac{7\pi}{34}](u^{1/5}+\sin[\frac{7\pi}{34}])\right]\cos[\frac{\pi}{34}]\\
&+ \frac{10}{17}\arctan\left[(u^{1/5}-\cos[\frac{2\pi}{17}])\csc[\frac{2\pi}{17}]\right]\cos[\frac{3\pi}{34}]
\end{aligned}
$$

$$
\begin{aligned}
&+\frac{10}{17}\arctan\left[\sec[\frac{\pi}{34}](u^{1/5}-\sin[\frac{\pi}{34}])\right]\cos[\frac{5\pi}{34}]\\
&+\frac{10}{17}\arctan\left[(u^{1/5}+\cos[\frac{\pi}{17}]\csc[\frac{\pi}{17}]\right]\cos[\frac{7\pi}{34}]\\
&-\frac{5}{17}\ln[-1+u^{1/5}]-\frac{5}{17}\cos[\frac{2\pi}{17}]\ln\left[1+u^{2/5}+2u^{1/5}\cos[\frac{3\pi}{17}]\right]\\
&+\frac{5}{17}\cos[\frac{3\pi}{17}]\ln\left[1+u^{2/5}-2u^{1/5}\cos[\frac{4\pi}{17}]\right]\\
&+\frac{5}{17}\cos[\frac{\pi}{17}]\ln\left[1+u^{2/5}+2u^{1/5}\sin[\frac{3\pi}{34}]\right]\\
&-\frac{5}{17}\cos[\frac{4\pi}{17}]\ln\left[1+u^{2/5}-2u^{1/5}\sin[\frac{5\pi}{34}]\right]\\
&-\frac{5}{17}\ln\left[1+u^{2/5}+2u^{1/5}\sin[\frac{7\pi}{34}]\right]\sin[\frac{\pi}{34}]\\
&+\frac{10}{17}\arctan\left[\sec[\frac{3\pi}{34}](u^{1/5}+\sin[\frac{3\pi}{34}])\right]\sin\left[\frac{\pi}{17}\right]\\
&+\frac{5}{17}\ln\left[1+u^{2/5}-2u^{1/5}\cos[\frac{2\pi}{17}]\right]\sin\left[\frac{3\pi}{34}\right]\\
&+\frac{10}{17}\arctan\left[(u^{1/5}+\cos[\frac{3\pi}{17}])\csc[\frac{3\pi}{17}]\right]\sin[\frac{2\pi}{17}]\\
&-\frac{5}{17}\ln\left[1+u^{2/5}-2u^{1/5}\sin[\frac{\pi}{34}]\right]\sin[\frac{5\pi}{34}]\\
&-\frac{10}{17}\arctan\left[(u^{1/5}-\cos[\frac{4\pi}{17}])\csc[\frac{4\pi}{17}]\right]\sin[\frac{3\pi}{17}]\\
&+\frac{5}{17}\ln\left[1+u^{2/5}+2u^{1/5}\cos[\frac{\pi}{17}]\right]\sin[\frac{7\pi}{34}]\\
&-\frac{10}{17}\arctan\left[\sec[\frac{5\pi}{34}](u^{1/5}-\sin[\frac{5\pi}{34}])\right]\sin[\frac{4\pi}{17}].
\end{aligned}
$$

(5) $$\int_0^u \frac{1}{1-u^{11/3}}du \quad (\text{for } N=11/3) \tag{D.5}$$

$$
\begin{aligned}
&=\frac{6}{11}\arctan\left[(u^{1/3}-\cos[\frac{2\pi}{11}])\csc[\frac{2\pi}{11}]\right]\cos[\frac{\pi}{22}]\\
&-\frac{6}{11}\arctan\left[(u^{1/3}+\sin[\frac{\pi}{22}])\sec[\frac{\pi}{22}]\right]\cos[\frac{3\pi}{22}]\\
&+\frac{6}{11}\arctan\left[(u^{1/3}+\cos[\frac{\pi}{11}])\csc[\frac{\pi}{11}]\right]\cos[\frac{5\pi}{22}]\\
&-\frac{3}{11}\ln[-1+u^{1/3}]+\frac{3}{11}\cos[\frac{\pi}{11}]\ln\left[1+u^{2/3}-2u^{1/3}\sin[\frac{3\pi}{22}]\right]
\end{aligned}
$$

$$
\begin{aligned}
&-\frac{3}{11}\cos[\frac{2\pi}{11}]\ln\left[1+u^{2/3}+2u^{1/3}\sin[\frac{5\pi}{22}]\right]\\
&+\frac{3}{11}\sin[\frac{\pi}{22}]\ln\left[1+u^{2/3}-2u^{1/3}\cos[\frac{2\pi}{11}]\right]\\
&-\frac{6}{11}\arctan\left[(u^{1/3}-\sin[\frac{3\pi}{22}])\sec[\frac{3\pi}{22}]\right]\sin[\frac{\pi}{11}]\\
&-\frac{3}{11}\sin[\frac{3\pi}{22}]\ln\left[1+u^{2/3}+2u^{1/3}\sin[\frac{\pi}{22}]\right]\\
&+\frac{6}{11}\arctan\left[(u^{1/3}+\sin[\frac{5\pi}{22}])\sec[\frac{5\pi}{22}]\right]\sin[\frac{2\pi}{11}]\\
&+\frac{3}{11}\sin[\frac{5\pi}{22}]\ln\left[1+u^{2/3}+2u^{1/3}\cos[\frac{\pi}{11}]\right].
\end{aligned}
$$

D.2 ETF-Based Solutions of the Second Integral for Five N-Values

The alternative analytical solutions of the second integral appearing in Eq. (3.12) for $N = 3, 10/3, 17/5, 7/2$ and $11/3$ for fully rough flows in wide channels ($M = 3$) can be expressed in terms of elementary transcendental functions (ETF), respectively.

(1) For $M = N = 3$

An equation for the ETF-based solutions of the second integral appearing in (3.12) for $M = N = 3$ is counted identically as (D.1) because the expressions of the first and second integrals appearing in (3.12) are identical if $M = N = 3$.

The remaining equations as shown below, i.e., (D.6) through (D.9), represent the ETF-based solutions of the second integral for N =10/3, 17/5, 7/2 and 11/3, respectively.

$$
\begin{aligned}
(2)\quad &\int_0^u \frac{u^{1/3}}{1-u^{10/3}}du \quad (\text{for } M=3 \text{ and } N=10/3) \qquad \text{(D.6)}\\
&=\frac{3}{40}\Bigg\{-2\sqrt{10-2\sqrt{5}}\arctan\left[\frac{1+\sqrt{5}-4u^{1/3}}{\sqrt{10-2\sqrt{5}}}\right]\\
&-2\sqrt{2(5+\sqrt{5})}\arctan\left[\frac{1-\sqrt{5}+4u^{1/3}}{\sqrt{2(5+\sqrt{5})}}\right]\\
&+2\sqrt{2(5+\sqrt{5})}\arctan\left[\frac{-1+\sqrt{5}+4u^{1/3}}{\sqrt{2(5+\sqrt{5})}}\right]\\
&-2\sqrt{10-2\sqrt{5}}\arctan\left[\frac{1+\sqrt{5}+4u^{1/3}}{\sqrt{10-2\sqrt{5}}}\right]
\end{aligned}
$$

$$
\begin{aligned}
&- 4\ln[-1+u^{1/3}] - 4\ln[1+u^{1/3}] \\
&- (-1+\sqrt{5})\ln\left[1-\frac{1}{2}(-1+\sqrt{5})u^{1/3}+u^{2/3}\right] \\
&- (-1+\sqrt{5})\ln\left[1+\frac{1}{2}(-1+\sqrt{5})u^{1/3}+u^{2/3}\right] \\
&+ (1+\sqrt{5})\ln\left[1-\frac{1}{2}(1+\sqrt{5})u^{1/3}+u^{2/3}\right] \\
&+ (1+\sqrt{5})\ln\left[1+\frac{1}{2}(1+\sqrt{5})u^{1/3}+u^{2/3}\right]\Bigg\}.
\end{aligned}
$$

$$
\begin{aligned}
(3)\ &\int_0^u \frac{u^{1/2}}{1-u^{7/2}}du \quad (\text{for } M = 3 \text{ and } N = 7/2) \qquad \text{(D.7)} \\
&= \frac{4}{7}\arctan\left[\csc[\frac{\pi}{7}](\sqrt{u}+\cos[\frac{\pi}{7}])\right]\cos[\frac{\pi}{14}] \\
&- \frac{4}{7}\arctan\left[\sec[\frac{\pi}{14}](\sqrt{u}+\sin[\frac{\pi}{14}])\right]\cos[\frac{3\pi}{14}] \\
&- \frac{2}{7}\ln[-1+\sqrt{u}] + \frac{2}{7}\cos[\frac{\pi}{7}]\ln\left[1+u-2\sqrt{u}\sin[\frac{3\pi}{14}]\right] \\
&+ \frac{2}{7}\sin[\frac{\pi}{14}]\ln\left[1+u+2\sqrt{u}\cos[\frac{\pi}{7}]\right] \\
&+ \frac{4}{7}\arctan\left[\sec[\frac{3\pi}{14}](\sqrt{u}-\sin[\frac{3\pi}{14}])\right]\sin[\frac{\pi}{7}] \\
&- \frac{2}{7}\sin[\frac{3\pi}{14}]\ln\left[1+u+2\sqrt{u}\sin[\frac{\pi}{14}]\right].
\end{aligned}
$$

$$
\begin{aligned}
(4)\ &\int_0^u \frac{u^{2/5}}{1-u^{17/5}}du \quad (\text{for } M = 3 \text{ and } N = 17/5) \qquad \text{(D.8)} \\
&= \frac{10}{17}\arctan\left[\sec[\frac{5\pi}{34}](u^{1/5}-\sin[\frac{5\pi}{34}])\right]\cos[\frac{\pi}{34}] \\
&+ \frac{10}{17}\arctan\left[(u^{1/5}+\cos[\frac{\pi}{17}])\csc[\frac{\pi}{17}]\right]\cos[\frac{3\pi}{34}] \\
&- \frac{10}{17}\arctan\left[\sec[\frac{\pi}{34}](u^{1/5}-\csc[\frac{4\pi}{17}])\right]\cos[\frac{4\pi}{17}] \\
&- \frac{10}{17}\arctan\left[(u^{1/5}-\sin[\frac{\pi}{34}])\sec[\frac{\pi}{34}]\right]\cos[\frac{7\pi}{34}] \\
&- \frac{5}{17}\ln[-1+u^{1/5}] + \frac{5}{17}\cos[\frac{3\pi}{17}]\ln\left[1+u^{2/5}-2u^{1/5}\cos[\frac{2\pi}{17}]\right]
\end{aligned}
$$

$$
\begin{aligned}
&-\frac{5}{17}\cos[\frac{4\pi}{17}]\ln\left[1+u^{2/5}+2u^{1/5}\cos[\frac{3\pi}{17}]\right]\\
&-\frac{5}{17}\cos[\frac{2\pi}{17}]\ln\left[1+u^{2/5}+2u^{1/5}\sin[\frac{3\pi}{34}]\right]\\
&+\frac{5}{17}\cos[\frac{\pi}{17}]\ln\left[1+u^{2/5}+2u^{1/5}\sin[\frac{7\pi}{34}]\right]\\
&-\frac{5}{17}\ln\left[1+u^{2/5}-2u^{1/5}\sin[\frac{5\pi}{34}]\right]\sin[\frac{\pi}{34}]\\
&+\frac{10}{17}\arctan\left[\sec[\frac{7\pi}{34}](u^{1/5}+\sin[\frac{7\pi}{34}])\right]\sin[\frac{\pi}{17}]\\
&+\frac{5}{17}\ln\left[1+u^{2/5}+2u^{1/5}\cos[\frac{\pi}{17}]\right]\sin[\frac{3\pi}{34}]\\
&+\frac{10}{17}\arctan\left[(u^{1/5}+\sin[\frac{3\pi}{34}])\sec[\frac{3\pi}{34}]\right]\sin[\frac{2\pi}{17}]\\
&-\frac{5}{17}\ln\left[1+u^{2/5}-2u^{1/5}\cos[\frac{4\pi}{17}]\right]\sin[\frac{5\pi}{34}]\\
&+\frac{10}{17}\arctan\left[(u^{1/5}-\cos[\frac{2\pi}{17}])\csc[\frac{2\pi}{17}]\right]\sin[\frac{3\pi}{17}]\\
&+\frac{5}{17}\ln[1+u^{2/5}-2u^{1/5}\sin[\frac{\pi}{34}]]\sin[\frac{7\pi}{34}]\\
&-\frac{10}{17}\arctan\left[\csc[\frac{3\pi}{17}](u^{1/5}+\cos[\frac{3\pi}{17}])\right]\sin[\frac{4\pi}{17}].
\end{aligned}
$$

(5)
$$
\int_0^u \frac{u^{2/3}}{1-u^{11/3}}du \quad (\text{for } M=3 \text{ and } N=11/3) \tag{D.9}
$$

$$
\begin{aligned}
&=\frac{6}{11}\arctan\left[(u^{1/3}+\cos[\frac{\pi}{11}])\csc[\frac{\pi}{11}]\right]\cos[\frac{\pi}{22}]\\
&-\frac{6}{11}\arctan\left[(u^{1/3}+\sin[\frac{5\pi}{22}])\sec[\frac{5\pi}{22}]\right]\cos[\frac{3\pi}{22}]\\
&+\frac{6}{11}\arctan\left[(u^{1/3}+\sin[\frac{\pi}{22}])\sec[\frac{\pi}{22}]\right]\cos[\frac{5\pi}{22}]\\
&-\frac{3}{11}\ln[-1+u^{1/3}]+\frac{3}{11}\cos[\frac{\pi}{11}]\ln\left[1+u^{2/3}-2u^{1/3}\cos[\frac{2\pi}{11}]\right]\\
&-\frac{3}{11}\cos[\frac{2\pi}{11}]\ln\left[1+u^{2/3}-2u^{1/3}\sin[\frac{3\pi}{22}]\right]\\
&+\frac{3}{11}\sin[\frac{\pi}{22}]\ln\left[1+u^{2/3}+2u^{1/3}\cos[\frac{\pi}{11}]\right]\\
&+\frac{6}{11}\arctan\left[(u^{1/3}-\cos[\frac{2\pi}{11}])\csc[\frac{2\pi}{11}]\right]\sin[\frac{\pi}{11}]
\end{aligned}
$$

$$
\begin{aligned}
&-\frac{3}{11}\sin[\frac{3\pi}{22}]\ln\left[1+u^{2/3}+2u^{1/3}\sin[\frac{5\pi}{22}]\right]\\
&-\frac{6}{11}\arctan\left[(u^{1/3}-\sin[\frac{3\pi}{22}])\sec[\frac{3\pi}{22}]\right]\sin[\frac{2\pi}{11}]\\
&+\frac{3}{11}\sin[\frac{5\pi}{22}]\ln\left[1+u^{2/3}+2u^{1/3}\sin[\frac{\pi}{22}]\right]
\end{aligned}
$$

Appendix E
Proof of the Identity Between the GHF-Based Solution and the Bresse Solution

We start with the GHF-based solutions of (3.8) in Chap. 3 in the domains of $| u | < 1$ and $| u | > 1$, as shown in (3.30) and (3.43), respectively, as well as with its corresponding ETF-based solution for the specified condition $M = N = 3$ (usually named Bresse solution), as shown in (3.78),

$$x_* = u \left[1 - g \left(\frac{1}{N}, u^N \right) \right] + \frac{\lambda^M u^{N-M+1}}{N-M+1} g \left(\frac{N-M+1}{N}, u^N \right) + \text{Const.}, \tag{3.30}$$

$$x_* = u \left[1 - \frac{u^{-N}}{N-1} g \left(\frac{N-1}{N}, u^{-N} \right) \right] + \frac{\lambda^M u^{-M+1}}{M-1} g \left(\frac{M-1}{N}, u^{-N} \right) + \text{Const.}, \tag{3.43}$$

$$x_* = u - \left[1 - \left(\frac{h_c}{h_n} \right)^3 \right] \times \left[\frac{1}{6} \ln \frac{(u^2+u+1)}{(u-1)^2} + \frac{1}{\sqrt{3}} \arctan \frac{(2u+1)}{\sqrt{3}} \right] + \text{Const.}, \tag{3.78}$$

in which $\lambda = h_c / h_n$. To represent the complete GHF-based solution for $M = N = 3$, we opt to use a combination of (3.30) for $0 \le u < 1$ and (3.43) for $u > 1$ as typical of the nine possible GHF-based solutions of (3.8) for $0 \le u < \infty$ excluding at $u = 1$, as mentioned in Chap. 3. Therefore, (3.30) on substitution of $M = N = 3$ yields

$$x_* = u \left[1 - g \left(\frac{1}{3}, u^3 \right) \right] + \lambda^3 u g \left(\frac{1}{3}, u^3 \right) + \text{Const.}, \tag{E.1}$$

C.-D. Jan, *Gradually-varied Flow Profiles in Open Channels*,
Advances in Geophysical and Environmental Mechanics and Mathematics,
DOI: 10.1007/978-3-642-35242-3, © Springer-Verlag Berlin Heidelberg 2014

which is valid for $0 \le u < 1$. Likewise, (3.43) on substitution of $M = N = 3$ yields

$$x_* = u\left[1 - \frac{u^{-3}}{2} g\left(\frac{2}{3}, u^{-3}\right)\right] + \frac{\lambda^3 u^{-2}}{2} g\left(\frac{2}{3}, u^{-3}\right) + \text{Const.}, \tag{E.2}$$

which is valid for $u > 1$. Rearrangeing (E.1) and (E.2), we obtain the more concise expressions

$$x_* = u\left[1 - (1-\lambda^3) g\left(\frac{1}{3}, u^3\right)\right] + \text{Const.}, \tag{E.3}$$

$$x_* = u - \frac{(1-\lambda^3)u^{-2}}{2} g\left(\frac{2}{3}, u^{-3}\right) + \text{Const.}, \tag{E.4}$$

respectively. The GHF-based solutions for $M = N = 3$ consist of two parts, i.e., (E.3) for $0 \le u < 1$ and (E.4) for $u > 1$, while the corresponding analytical solution of (3.8) obtained from the Mathematica software expressed in terms of the ETF for $M = N = 3$ is (3.78), which has been shown to be derivable from (3.12) upon substitution of (D.1) from Appendix D.

E.1 Two Identity Relations Between the GHF-Based and ETF-Based Functions

Comparison of (E.3) for $0 \le u < 1$ and (E.4) for $u > 1$ with (3.78) for $0 \le u < \infty$, except $u = 1$, leads to the formulation of the following two identities:

$$u g\left(\frac{1}{3}, u^3\right) = \frac{1}{6}\ln\left(\frac{u^2+u+1}{(u-1)^2}\right) + \frac{1}{\sqrt{3}}\arctan\left(\frac{2u+1}{\sqrt{3}}\right) + \text{Const.}, \tag{E.5}$$

which is valid for $0 \le u < 1$, and

$$\frac{u^{-2}}{2} g\left(\frac{2}{3}, u^{-3}\right) = \frac{1}{6}\ln\left(\frac{u^2+u+1}{(u-1)^2}\right) + \frac{1}{\sqrt{3}}\arctan\left(\frac{2u+1}{\sqrt{3}}\right) + \text{Const.}, \tag{E.6}$$

which is valid for $u > 1$. The constant of integration, "Const." in (E.5) or (E.6) is determined using a boundary condition in the respective domain of u. We determine "Const." in (E.5) and (E.6) subject to the boundary conditions $u = 0$ and $u = \infty$, respectively, as follows: Substituting $u = 0$ into (E.5) yields

$$C_0 = -\frac{1}{\sqrt{3}}\arctan\left(\frac{1}{\sqrt{3}}\right) = -\frac{\pi}{6\sqrt{3}}, \tag{E.7}$$

in which $\arctan(1/\sqrt{3}) = \pi/6$ because $\arctan(1/\sqrt{3})$ is the principal value of the angle whose tangent is $1/\sqrt{3}$ and is defined in the Mathematica software as $-\pi/2 \leq \arctan(1/\sqrt{3}) \leq \pi/2$. Likewise, substituting $u = \infty$ into (E.6) yields

$$C_\infty = -\frac{1}{\sqrt{3}} \arctan(\infty) = -\frac{\pi}{2\sqrt{3}}, \tag{E.8}$$

in which $\arctan(\infty) = \pi/2$ because $\arctan(\infty)$ is the principal value of the angle whose tangent is ∞ within the afore-defined range in the Mathematica software. This is attributed to the fact that the logarithmic function appearing in the first term on the right-hand side of (E.6) disappears by virtue of L'Hopital's rule, namely

$$\lim_{u \to \infty} \ln\left(\frac{u^2 + u + 1}{(u-1)^2}\right) = 0. \tag{E.9}$$

It follows that (E.5) and (E.6) on substitution of "Const." by C_0 from (E.7) and by C_∞ from (E.8), respectively, yield

$$ug\left(\frac{1}{3}, u^3\right) = \frac{1}{6}\ln\left(\frac{u^2 + u + 1}{(u-1)^2}\right) + \frac{1}{\sqrt{3}}\arctan\left(\frac{2u+1}{\sqrt{3}}\right) - \frac{\pi}{6\sqrt{3}}, \tag{E.10}$$

$$\frac{u^{-2}}{2}g\left(\frac{2}{3}, u^{-3}\right) = \frac{1}{6}\ln\left(\frac{u^2 + u + 1}{(u-1)^2}\right) + \frac{1}{\sqrt{3}}\arctan\left(\frac{2u+1}{\sqrt{3}}\right) - \frac{\pi}{2\sqrt{3}}. \tag{E.11}$$

Equation (E.10) is an identity being valid for $0 \leq u < 1$ between the GHF and the corresponding logarithmic—inverse trigonometric functions, while (E.11) is another identity being valid for $u > 1$ between the GHF and the corresponding logarithmic—inverse trigonometric functions. It is noted that (E.10) and (E.11) are useful in the proof of the identity between the GHF-based solution for $M = N = 3$ and the Bresse solution. Two examples are given in the following to substantiate such identities.

E.2 Proof of the Identity Between the GHF-Based and ETF-Based Solutions

For example, to plot the M2, M3, and S2 profiles, we can use either the GHF-based solution [i.e., (E.3)] or the ETF-based solution [i.e., (3.78)], and it can be shown that both solution equations should yield identical results. A boundary condition $(x_*, u) = (0, u_1)$ for plotting one of such profiles is arbitrarily selected at $x_* = 0$, where the flow depth, h, is assumed to be $h_1 = u_1 h_n$ (u_1 = any positive real number). Upon substitution of this boundary condition, the "Const." terms in (E.3) and (3.78) can, respectively, be expressed

$$C_1 = -u_1 + (1-\lambda^3)u_1 g\left(\frac{1}{3}, u_1^3\right), \tag{E.12}$$

$$C_2 = -u_1 + (1-\lambda^3)\left[\frac{1}{6}\ln\left(\frac{u_1^2+u_1+1}{(u_1-1)^2}\right) + \frac{1}{\sqrt{3}}\arctan\left(\frac{2u_1+1}{\sqrt{3}}\right)\right]. \tag{E.13}$$

When we plot either (E.3) on substitution of "Const." by C_1 from (E.12) or (3.78) on substitution of "Const." by C_2 from (E.13), we find that both profiles so plotted must be identical because subtracting the right-hand expression of (3.78) from that of (E.3) or vice versa, the result can be proved to be zero. In other words, we have

$$\begin{aligned}(1-\lambda^3)&\left\{-ug\left(\frac{1}{3}, u^3\right) + \left[\frac{1}{6}\ln\left(\frac{u^2+u+1}{(u-1)^2}\right) + \frac{1}{\sqrt{3}}\arctan\left(\frac{2u+1}{\sqrt{3}}\right)\right]\right\}\\ + (1-\lambda^3)&\left\{u_1 g\left(\frac{1}{3}, u_1^3\right) - \left[\frac{1}{6}\ln\left(\frac{u_1^2+u_1+1}{(u_1-1)^2}\right)\right.\right.\\ &\left.\left. + \frac{1}{\sqrt{3}}\arctan\left(\frac{2u_1+1}{\sqrt{3}}\right)\right]\right\} = 0,\end{aligned} \tag{E.14}$$

because adding or subtracting the same constant, $-(1/\sqrt{3})\arctan(1/\sqrt{3})$, to or from the expressions appearing in the two square brackets of (E.14) can offset each other, but cause the expressions within the two braces to vanish by virtue of (E.7) and (E.10).

In another example, if the M1, S1, and S2 profiles are plotted, we can use either (E.4) or (3.78). Likewise, it can be readily shown that both equations should yield an identical result. A boundary condition $(x_*, u) = (0, u_2)$ for plotting one of such profiles is arbitrarily selected at $x_* = 0$, where the flow depth, h, is assumed to be $h_2 = u_2 h_n$ (u_2 = any positive real number). The "Const." in (E.4) and (3.78) can be respectively expressed from (E.4) and (3.78) upon substitution of this boundary condition as

$$C_3 = -u_2 + (1-\lambda^3)\frac{u_2^{-2}}{2} g\left(\frac{2}{3}, u_2^{-3}\right), \tag{E.15}$$

$$b4\, C_4 = -u_2 + (1-\lambda^3)\left[\frac{1}{6}\ln\left(\frac{u_2^2+u_2+1}{(u_2-1)^2}\right) + \frac{1}{\sqrt{3}}\arctan\left(\frac{2u_2+1}{\sqrt{3}}\right)\right]. \tag{E.16}$$

When we plot either (E.4) on substitution of "Const." by C_3 from (E.15) or (3.78) on substitution of "Const." by C_4 from (E.16), we find that both profiles so plotted should be identical because subtracting the right-hand expression of (3.78) from that of (E.4) or vice versa the result can again be proved to be zero. In other words, we have

$$
\begin{aligned}
&(1-\lambda^3)\left\{-\frac{u^{-2}}{2}g\left(\frac{2}{3},u^{-3}\right)\right.\\
&\qquad\left.+\left[\frac{1}{6}\ln\left(\frac{u^2+u+1}{(u-1)^2}\right)+\frac{1}{\sqrt{3}}\arctan\left(\frac{2u+1}{\sqrt{3}}\right)\right]\right\}\\
&+(1-\lambda^3)\left\{\frac{u_2^{-2}}{2}g\left(\frac{2}{3},u_2^{-3}\right)\right.\\
&\qquad\left.-\left[\frac{1}{6}\ln\left(\frac{u_2^2+u_2+1}{(u_2-1)^2}\right)+\frac{1}{\sqrt{3}}\arctan\left(\frac{2u_2+1}{\sqrt{3}}\right)\right]\right\}=0.
\end{aligned}
\tag{E.17}
$$

because adding or subtracting the very same constant, $-(1/\sqrt{3})\arctan(\infty)$, to or from the expressions appearing in the two square brackets of (E.17) can offset each other, but cause the expressions within the two braces to vanish by virtue of (E.8) and (E.11).

E.3 Alternative Forms of the Identity Relations and Bresse Solution

In Appendix E, we have analytically proved that the complete GHF-based solutions, i.e., a combination of (3.30) for $0 \le u < 1$ and (3.43) for $u > 1$, in a particular case of $M = N = 3$ is exactly identical to the so-called Bresse solution, i.e., (3.78), the simplest ETF-based solution of (3.8).

In addition, there is a relation between $\arctan z$ and $\operatorname{arccot} z$ for a variable z. That is to say,

$$
\arctan\left(\frac{2u+1}{\sqrt{3}}\right)+\operatorname{arccot}\left(\frac{2u+1}{\sqrt{3}}\right)=\frac{\pi}{2}. \tag{E.18}
$$

Therefore, (E.10) for $0 \le u < 1$ and (E.11) for $u > 1$, can be written as (E.19) and (E.20), respectively.

$$
ug\left(\frac{1}{3},u^3\right)=\frac{1}{6}\ln\left(\frac{u^2+u+1}{(u-1)^2}\right)-\frac{1}{\sqrt{3}}\operatorname{arccot}\left(\frac{2u+1}{\sqrt{3}}\right)+\frac{\pi}{3\sqrt{3}}, \tag{E.19}
$$

$$
\frac{u^{-2}}{2}g\left(\frac{2}{3},u^{-3}\right)=\frac{1}{6}\ln\left(\frac{u^2+u+1}{(u-1)^2}\right)-\frac{1}{\sqrt{3}}\operatorname{arccot}\left(\frac{2u+1}{\sqrt{3}}\right) \tag{E.20}
$$

and, the Bresse solution [i.e., (3.78)] can also be written in another form as

$$
x_*=u-(1-\lambda^3)\left[\frac{1}{6}\ln\left(\frac{u^2+u+1}{(u-1)^2}\right)-\frac{1}{\sqrt{3}}\operatorname{arccot}\left(\frac{2u+1}{\sqrt{3}}\right)\right]+\text{Const.}. \tag{E.21}
$$

Appendix F
Obtain the GHF-Based Solution for $\mid \lambda v \mid > 1$ from that for $\mid \lambda v \mid < 1$, or Vice Versa

Let us start with the GHF-based solutions of (4.4) in the domains of $\mid \lambda v \mid < 1$ and $\mid \lambda v \mid > 1$ for GVF in sustaining channels, viz.,

$$x_\sharp = \frac{v^{N-M+1}}{N-M+1} g\left(\frac{N-M+1}{N}, (\lambda v)^N\right) - \frac{v^{N+1}}{N+1} g\left(\frac{N+1}{N}, (\lambda v)^N\right) + \text{Const.}, \tag{4.32}$$

$$x_\sharp = \lambda^{-N} v g\left(-\frac{1}{N}, (\lambda v)^{-N}\right) + \frac{\lambda^{-N} v^{-M+1}}{M-1} g\left(\frac{M-1}{N}, (\lambda v)^{-N}\right) + \text{Const.} \tag{4.47}$$

and, the GHF-based solutions of (4.12) in the domains of $\mid \lambda v \mid < 1$ and $\mid \lambda v \mid > 1$ for GVF in adverse channels,

$$x_\sharp = \frac{v^{N-M+1}}{N-M+1} g\left(\frac{N-M+1}{N}, -(\lambda v)^N\right) - \frac{v^{N+1}}{N+1} g\left(\frac{N+1}{N}, -(\lambda v)^N\right) + \text{Const.}, \tag{4.69}$$

$$x_\sharp = -\lambda^{-N} v g\left(-\frac{1}{N}, -(\lambda v)^{-N}\right) - \frac{\lambda^{-N} v^{-M+1}}{M-1} g\left(\frac{M-1}{N}, -(\lambda v)^{-N}\right) + \text{Const.}. \tag{4.84}$$

C.-D. Jan, *Gradually-varied Flow Profiles in Open Channels*,
Advances in Geophysical and Environmental Mechanics and Mathematics,
DOI: 10.1007/978-3-642-35242-3, © Springer-Verlag Berlin Heidelberg 2014

For specified parameter relations, $a = 1$ and $c = b + 1$, the relation connecting one GHF in the domain of $|\, z \,| < 1$ to two GHF in the domain of $|\, z \,| > 1$, as worked out in detail in Appendices A and C, can be written as

$$\begin{aligned} &{}_2F_1(1, b; 1 + b; z) \\ &\quad = \tfrac{b}{b-1}(-z^{-1})\, {}_2F_1(1, 1 - b; 2 - b; z^{-1}) + \Gamma(1 - b)\Gamma(1 + b)\left(-z^{-1}\right)^b . \end{aligned} \tag{F.1}$$

The second term on the right-hand side of (F.1) has been simplified from the second GHF of the relation in the domain of $|\, z \,| > 1$. For reducing the related equations to shorter expressions in order to facilitate the reading of the manuscript, we will use $g(b, z)$ instead of ${}_2F_1(1, b; b + 1; z)$ herein to represent the GHF when we treat the GHF-solutions. Therefore, the shorter expression of (F.1) is given as

$$g(b; z) = \frac{b}{b-1}(-z^{-1})g(1 - b, z^{-1}) + \Gamma(1 - b)\Gamma(1 + b)(-z^{-1})^b. \tag{F.2}$$

We exemplify (F.1) or (F.2) so formulated for the following two cases. The first case is to derive the GHF-based solution of (4.4) in the domain of $|\, \lambda v \,| > 1$ from its counterpart in the domain $|\, \lambda v \,| < 1$ for GVF in sustaining channels, and the second case is to derive the GHF-based solution of (4.12) in the domain of $|\, \lambda v \,| > 1$ from its counterpart in the domain $|\, \lambda v \,| < 1$ for GVF in adverse channels. In what follows, we work out both cases using (F.2) without recourse to the formulation of (4.13) and (4.14) [i.e., alternative forms of (4.4) and (4.12), respectively, with their variable, w, being expressed as v^{-1}], which can be solved for their respective GHF-based GVF profiles in the domain of $|\, \lambda v \,| > 1$.

F.1 Obtain (4.47) Directly from (4.32) by Using the Transformation Relation

Because the first and second GHF on the right-hand side of (4.32) have the specified parameter relations, $b = (N + M - 1)/N$ and $b = (N + 1)/N$, respectively, (F.2) on substitution of such b-expressions to the first and second GHF on the right-hand side of (4.32) yield, respectively,

$$\begin{aligned} g\left(\frac{N - M + 1}{N}, (\lambda v)^N\right) = {} & \frac{(N - M + 1)(\lambda v)^{-N}}{M - 1} g\left(\frac{M - 1}{N}, (\lambda v)^{-N}\right) \\ & + \Gamma\left(\frac{M - 1}{N}\right)\Gamma\left(\frac{2N - M + 1}{N}\right) \\ & - \left(-(\lambda v)^{-N}\right)^{(N - M + 1)/N} , \end{aligned} \tag{F.3}$$

$$g\left(\frac{N+1}{N}, (\lambda v)^N\right) = -\frac{(\lambda v)^{-N}}{N+1} g\left(\frac{-1}{N}, (\lambda v)^{-N}\right) + \Gamma\left(\frac{N-1}{N}\right)\Gamma\left(\frac{N+1}{N}\right)\left(-(\lambda v)^{-N}\right)^{1/N}. \tag{F.4}$$

Substituting (F.3) and (F.4), respectively, into the first and second GHF appearing on the right-hand side of (4.32), yields

$$\begin{aligned} x_{\sharp} = {} & \lambda^{-N} v g\left(-\frac{1}{N}, (\lambda v)^{-N}\right) + \frac{\lambda^{-N} v^{-M+1}}{M-1} g\left(\frac{M-1}{N}, (\lambda v)^{-N}\right) \\ & + \frac{(-1)^{(N+1)/N} \lambda^{-(N+1)}}{N+1} \Gamma\left(-\frac{1}{N}\right)\Gamma\left(\frac{2N+1}{N}\right) \\ & + \frac{(-1)^{(N-M+1)/N} \lambda^{-(N-M+1)}}{N-M+1} \Gamma\left(\frac{M-1}{N}\right)\Gamma\left(\frac{2N-M+1}{N}\right) + \text{Const.}. \end{aligned} \tag{F.5}$$

Comparison of (F.5) with (4.47) reveals that (F.5) is identical to (4.47) because two constant terms appearing in (F.5) in addition to "Const." can be ignored without losing the general meaning of the constant term. Consequently, one may conclude that (4.47) obtained from (4.13) is indeed the GHF-based solution of (4.4) in the domain of $\mid \lambda v \mid > 1$. This reconfirms the validity of (4.47) in the domain of $\mid \lambda v \mid > 1$ and the uniqueness of a role of (4.47) playing in the computation of GVF profiles in sustaining channels.

F.2 Obtain (4.84) Directly from (4.69) by Using the Transformation Relation

Likewise, in the second case, we use (F.2) to transform the GHF-based solution of (4.12) in the domain of $\mid \lambda v \mid < 1$, i.e., (4.69), to the corresponding GHF-based solution of (4.12) in the domain of $\mid (h_c/h_n)v \mid > 1$, i.e., (4.84) for GVF in adverse channels. In the following, we can prove that (4.69) for $\mid \lambda v \mid < 1$ is transformable to (4.84) for $\mid \lambda v \mid > 1$, which we have actually obtained from (4.14) [i.e., an alternative form of (4.12) with its variable, w, being expressed as v^{-1}] rather than through (4.12).

Again, we have $b = (N - M + 1)/N$ and $b = (N + 1)/N$ in the first and second GHF, respectively, with the variable $z = -(\lambda v)^N$ of both GHF appearing on the right-hand side of (4.69); therefore, (F.2) on substitution of such two respective b-expressions yield, respectively,

$$g\left(\frac{N-M+1}{N}, -(\lambda v)^N\right) = \frac{(N-M+1)(\lambda v)^{-N}}{-M+1} g\left(\frac{M-1}{N}, -(\lambda v)^{-N}\right) + \Gamma\left(\frac{M-1}{N}\right)\Gamma\left(\frac{2N-M+1}{N}\right)(\lambda v)^{-(N-M+1)}, \tag{F.6}$$

$$g\left(\frac{N+1}{N}, -(\lambda v)^N\right) = (\lambda v)^{-N}(N+1)g\left(\frac{-1}{N}, -(\lambda v)^{-N}\right) + \Gamma\left(\frac{-1}{N}\right)\Gamma\left(\frac{2N+1}{N}\right)(\lambda v)^{-(N+1)}. \tag{F.7}$$

Substituting (F.6) and (F.7) into the first and second GHF appearing on the right-hand side of (4.69), respectively, yields

$$\begin{aligned} x_\sharp = &-\lambda^{-N} v g\left(-\frac{1}{N}, -(\lambda v)^{-N}\right) - \frac{\lambda^{-N} v^{-M+1}}{M-1} g\left(\frac{M-1}{N}, -(\lambda v)^{-N}\right) \\ &+ \frac{\lambda^{-(N+1)}}{N+1}\Gamma\left(-\frac{1}{N}\right)\Gamma\left(\frac{2N+1}{N}\right) \\ &+ \frac{\lambda^{-(N-M+1)}}{N-M+1}\Gamma\left(\frac{M-1}{N}\right)\Gamma\left(\frac{2N-M+1}{N}\right) + \text{Const.}. \end{aligned} \tag{F.8}$$

Comparison of (F.8) with (4.84) reveals that (F.8) is identical to (4.84) because two constant terms appearing in (F.8) in addition to "Const." can be ignored without losing the gereral meaning of the constant term.

Appendix G
Asymptotic Reduction of the GHF-Based GVF Solutions for Flow in Sustaining and Adverse Channels to the Solutions for Flow in Horizontal Channels

The starting point is the GHF-based solutions of (4.4) in the domain $\mid \lambda v \mid < 1$ for GVF in sustaining channels:

$$x_{\sharp} = \frac{v^{N-M+1}}{N-M+1} g\left(\frac{N-M+1}{N}, (\lambda v)^N\right) - \frac{v^{N+1}}{N+1} g\left(\frac{N+1}{N}, (\lambda v)^N\right) + \text{Const.} \tag{4.32}$$

and, the GHF-based solutions of (4.12) in the domain $0 \leq \lambda v < \infty$ for GVF in adverse channels:

$$x_{\sharp} = \frac{v^{N-M+1}}{N-M+1} g\left(\frac{N-M+1}{N}, -(\lambda v)^N\right) - \frac{v^{N+1}}{N+1} g\left(\frac{N+1}{N}, -(\lambda v)^N\right) + \text{Const.}. \tag{4.69}$$

To prove that GVF profiles on the horizontal slope can also be derived from the asymptotic reduction of their counterparts on the sustaining slope ($\theta > 0$) or on the adverse slope ($\theta < 0$) as $\theta \to 0$, we should simultaneously examine: firstly, whether the valid domain of v for each GHF-based solution of (4.4) and (4.12) can reduce asymptotically to that of (4.11); and secondly, whether each GHF-based solution of (4.4) and (4.12) can by itself reduce asymptotically to (4.11) as λ ($= h_c/h_n$) $\to 0$ (or $h_n \to \infty$). As shown in Table 5.1, the successful completion of one of the above two analyses without support from the other does not necessarily warrant the asymptotic reduction of GVF profiles on the sustaining or adverse slope to their counterparts on the horizontal slope as $\theta \to 0$. It appears simpler to examine the asymptotic reduction of the valid domain of v for each GHF-based solution of (4.4) and (4.12) to that for (4.11) as $\lambda \to 0$ because we can readily grasp it without elaborating its proof. However, for clarity, our attention should be more focused below on the scrutiny of each GHF-based solution of (4.4) and (4.12) by checking the existence of

C.-D. Jan, *Gradually-varied Flow Profiles in Open Channels*,
Advances in Geophysical and Environmental Mechanics and Mathematics,
DOI: 10.1007/978-3-642-35242-3, © Springer-Verlag Berlin Heidelberg 2014

its limit at $\lambda = 0$. This is in fact more critical than simply checking its valid domain of v, thus enabling us to determine which one of the GHF-based solutions of (4.4) and (4.12) can reduce asymptotically to (4.11) as $\lambda \to 0$.

The proof of the asymptotic reduction of each GHF-based solution of (4.4) and (4.12) to (4.11) as $\lambda \to 0$ (or $h_n \to \infty$) can be handled by substituting $\lambda = 0$ into every term in the hypergeometric series, as defined in (A.5) of Appendix A. For the convenience of discussion, we rewrite the definition of GHF [i.e., (A.5)] here as

$$
{}_2F_1(a, b; c; z) = \frac{\Gamma(c)}{\Gamma(a)\Gamma(b)} \sum_{k=0}^{\infty} \frac{\Gamma(a+k)\Gamma(b+k)}{\Gamma(c+k)k!} z^k, \tag{G.1}
$$

in which $| z | < 1$. The expansion of the hypergeometric series is given as

$$
\begin{aligned}
&{}_2F_1(a, b; c; z) \\
&\quad = \frac{\Gamma(c)}{\Gamma(a)\Gamma(b)} \left\{ \frac{\Gamma(a)\Gamma(b)}{\Gamma(c)0!} + \frac{\Gamma(a+1)\Gamma(b+1)}{\Gamma(c+1)1!} z + \ldots \right\} \\
&\quad = \left\{ 1 + \frac{ab}{c} z + \frac{ab(a+1)(b+1)}{c(c+1)2!} z^2 \ldots \right\}.
\end{aligned} \tag{G.2}
$$

Since $z^0 = 1$ and $0! = 1$, the first term of GHF in (G.1) is unity as shown in (G.2). The second or higher term ($k \geq 1$) of every GHF vary with the expressions of the parameters a, b, and c as well as the form of the variable z in ${}_2F_1(a, b; c; z)$. Since every GHF appearing in each GHF-based solution of (4.4) and (4.12) has the property of $a = 1$ and $c = b + 1$, the simpler expression $g(b, z)$ is used to replace ${}_2F_1(a, b; c; z)$ for shortening the related expressions,

$$
g(b, z) = {}_2F_1(1, b; b+1; z) = b \sum_{k=0}^{\infty} \frac{z^k}{b+k} = 1 + \frac{bz}{b+1} + \frac{bz^2}{b+2} + \frac{bz^3}{b+3} + \ldots. \tag{G.3}
$$

As $\lambda\,(= h_c/h_n) \to 0$, we can express the asymptotic reduction of the first and second terms on the right-hand side of (4.32), respectively, as

$$
\begin{aligned}
&\lim_{\lambda \to 0} \left\{ \frac{v^{N-M+1}}{N-M+1} g\left(\frac{N-M+1}{N}, (\lambda v)^N \right) \right\} \\
&\qquad = \lim_{\lambda \to 0} \left\{ \frac{v^{N-M+1}}{N-M+1} \left(1 + \frac{(N-M+1)(\lambda v)^N}{2N-M+1} + \ldots \right) \right\} \\
&\qquad = \frac{v^{N-M+1}}{N-M+1},
\end{aligned} \tag{G.4}
$$

$$\lim_{\lambda\to 0}\left\{-\frac{\upsilon^{N+1}}{N+1}g\left(\frac{N+1}{N},(\lambda\upsilon)^N\right)\right\}$$
$$=\lim_{\lambda\to 0}\left\{-\frac{\upsilon^{N+1}}{N+1}\left[1+\frac{(N+1)(\lambda\upsilon)^N}{2N+1}+...\right]\right\} \tag{G.5}$$
$$=-\frac{\upsilon^{N+1}}{N+1}.$$

Adding (G.4) and (G.5) yields exactly the right-hand expression of (4.11), thus proving that (4.32) reduces asymptotically to (4.11) as λ ($= h_c/h_n$) $\rightarrow 0$.

Clearly, if and only if (4.32) is used to describe M2 and M3 profiles on the M slope ($0 < h_c/h_n < 1$), it can be proved to have the limit at λ ($= h_c/h_n$) $\rightarrow 0$, as shown in (G.4) and (G.5). In fact, as $h_c/h_n \rightarrow 0$, the valid domain of υ for M2 profiles reduces from $1 \leq \upsilon < h_n/h_c$ to $1 \leq \upsilon < \infty$, which is identical to that for the H2 profile, and left intact is the valid domain of υ for M3 profiles, i.e., $0 \leq \upsilon \leq 1$, which is identical to that for the H3 profile.

The asymptotic reduction of (4.69) to (3.11) as $h_c/h_n \rightarrow 0$ can be treated in the same way as that of (4.32) to (3.11). In the treatment, we pay attention to the sole difference in the expressions of the two corresponding GHF appearing in (4.32) and (4.69). Comparing them, we can readily see that it is the positive or negative sign of the variable in one set of the two GHF that differs from the other set of the two GHF; otherwise, the two corresponding GHF shown in (4.32) and (4.69) are identical. Following the same evaluation procedure which has resulted in (G.4) and (G.5), we can acquire that the first and second terms on the right-hand side of (4.69) reduce asymptotically to the following respective expressions as $h_c/h_n \rightarrow 0$.

As λ ($= h_c/h_n$) $\rightarrow 0$, we can express the asymptotic reduction of the first and second terms on the right-hand side of (4.69), respectively, as

$$\lim_{\lambda\to 0}\left\{\frac{\upsilon^{N-M+1}}{N-M+1}g\left(\frac{N-M+1}{N},-(\lambda\upsilon)^N\right)\right\}$$
$$=\lim_{\lambda\to 0}\left\{\frac{\upsilon^{N-M+1}}{N-M+1}\left(1-\frac{(N-M+1)(\lambda\upsilon)^N}{2N-M+1}+...\right)\right\} \tag{G.6}$$
$$=\frac{\upsilon^{N-M+1}}{N-M+1},$$

$$\lim_{\lambda\to 0}\left\{-\frac{\upsilon^{N+1}}{N+1}g\left(\frac{N+1}{N},-(\lambda\upsilon)^N\right)\right\}$$
$$=\lim\nolimits_{\lambda\to 0}\left\{-\frac{\upsilon^{N+1}}{N+1}\left[1-\frac{(N+1)(\lambda\upsilon)^N}{2N+1}+...\right]\right\} \tag{G.7}$$
$$=-\frac{\upsilon^{N+1}}{N+1}.$$

Adding (G.6) and (G.7) yields the right-hand expression of (4.11) too, thus proving that (4.69) also reduces asymptotically to (4.11) as $\lambda(= h_c/h_n) \rightarrow 0$. In fact, as $\lambda \rightarrow 0$, the valid domain of υ for A2 profiles reduces from $1 \leq \upsilon < h_n/h_c$ to

$1 \leq \upsilon < \infty$, which is identical to that for the H2 profile, and left intact is the valid domain of υ for A3 profiles, i.e., $0 \leq \upsilon \leq 1$, which is identical to that for the H3 profile. These results are summarized in Table 5.1 in Chap. 5.

References

Abramowitz M, Stegun IA (1972) Handbook of mathematical functions with formulas, graphs, and mathemtical tabels,10th edn. Number 55 in Applied mathematics series. National Buereau of Standards, Department of Commerce, USA.
Bernardin L, Chin P, DeMarco P, Geddes KO, Hare DEG, Heal KM, Labahn G, May JP, McCarron J, Monagan MB, Ohashi D, Vorkoetter SM (2011) Maple programming guide. Maplesoft, Waterloo Maple Inc., Waterloo.
Jan CD, Chen CL (2012) Use of the Gaussian hypergeometric function to solve the equation of gradually-varied flow. J Hydrol 456–457:139–145
Jan CD, Chen CL (2013), Gradually varied open-channel flow profiles normalized by critical depth and analytically solved by using gaussian hypergeometric functions. Hydrol Earth, Syst Sci (HESS)17:973–987.
Korn GA, Korn TM (1961) Mathematical handbook for scientists and engineers. McGraw-Hill, New York
Kummer E (1836) Über die hypergeometrische Reihe. J für die Reine und Angewandte Mathematik (15):39–83, 127–172.
Luke YL (1975) Mathematical functions and their approximations. Academic Press Inc., New York
Pearson CE (1974) Handbook of applied mathematics. Selected Results and Methods, Van Nostrand Reinhold Company, New York
Seaborn JB (1991) Hypergeometric functions and their applications. Springer, New York
Whittaker ET, Watson GT (1992) An introduction to the general theory of infinite processes and of analytic functions with an account of the principal transcendental functions. Cambridge University Press, New York, A course of modern analysis
Wolfram S (1996) The mathematica book, 3rd edn. Wolfram Media and Cambridge University Press, Champaign

Author Index

C.-D. Jan, *Gradually-varied Flow Profiles in Open Channels*,
Advances in Geophysical and Environmental Mechanics and Mathematics,
DOI: 10.1007/978-3-642-35242-3,

© Springer-Verlag Berlin Heidelberg 2014

Subject Index

C.-D. Jan, *Gradually-varied Flow Profiles in Open Channels*,
Advances in Geophysical and Environmental Mechanics and Mathematics,
DOI: 10.1007/978-3-642-35242-3, © Springer-Verlag Berlin Heidelberg 2014

MIX
Papier aus verantwortungsvollen Quellen
Paper from responsible sources
FSC® C105338

If you have any concerns about our products,
you can contact us on
ProductSafety@springernature.com

In case Publisher is established outside the EU,
the EU authorized representative is:
Springer Nature Customer Service Center GmbH
Europaplatz 3, 69115 Heidelberg, Germany

Printed by Libri Plureos GmbH
in Hamburg, Germany